The Great Climate Change Debate
Karoly v Happer

By Andy May

Hardback Version: 978-1-63944-674-2
Paperback Version: 978-1-63944-676-6
eBook Version: 978-1-63944-678-0

Manuscript Editor: Robert Burger

First Edition-

Printed in the United States of America

Dedication

To debate and critical thinking.

Other Works by The Author

Climate Catastrophe! Science or Science Fiction?
Blood and Honor: The People of Bleeding Kansas
Politics and Climate Change: A History

Acknowledgements

To my wife Aurelia for her patience during the preparation of another book by her husband. To Dr. Martin Capages Jr., the publisher of my first three books and who suggested I write this book. Capages, David Siegel, and James Barham reviewed an early draft of the book and made many helpful suggestions. I also acknowledge the help of my copyeditor, Robert Burger, who greatly improved the book. Finally, I would like to thank my friend Dr. Javier Vinós who carefully reviewed a near final version of the book.

Table of Contents

Foreword

The scientific method now appears to be an obsolete concept, in the time of computer models smartly programmed to reproduce the ignorance of their coders. Yet after being rigorously trained in it for years, I conducted scientific research for several decades in molecular genetics, neurobiology, and cancer at some of the most prestigious research institutions, including the Howard Hughes Medical Institute at the University of California at San Diego, and the famous Laboratory of Molecular Biology of the Medical Research Council in Cambridge, UK. At those places I was fortunate to meet some of the most intelligent people in the planet, including Nobel Prize recipients Roger Tsien, Max Perutz, and César Milstein, who in seminars and informal talks furthered my scientific education.

As a biologist, I was very aware of global warming since the mid-1980s, and I remember the cold winters of my childhood. As a scientist, I had no reason to doubt the conclusions of climate scientists without checking their evidence. It was not until late 2014 when I had the time and inclination to do so. Since scientist's opinions do not constitute science, what I do is read the scientific articles to see if the conclusions reached are fully supported by the evidence presented and look for alternative explanations. At first, I was surprised by the contrast between the weakness of the evidence and the confidence in the conclusions. Then I became worried because not only were natural explanations neglected, but vital environmental defense was diluted and forgotten due to the fight against global warming. Finally, I became horrified at the suppression of dissent and the mistreatment of skeptical scientists. This was not the truth-seeking science that has served humankind so well in the past.

Since I am not a climate scientist, I started writing articles in climate-science blogs, bringing to wider audiences, peer-reviewed climate evidence that supports alternative explanations for modern global warming. Particularly I investigated paleoclimatology, looking for examples of past climate changes due to natural causes, even in the absence of significant greenhouse gas changes. The idea that natural climate change stopped the moment humans developed an industrial civilization, as promoted by the IPCC, is an insult to our intelligence.

I got to know Andy after publishing two articles on the role of solar variability on climate change during the Holocene at different blogs. He wrote a review article highlighting my main findings at the "Watts Up With That?" (wattsupwiththat.com) blog in September 2016. The final phrase in that article was: "English is not Javier's first language and we need to look past this, but his research and content are first rate." In a comment to the article, I thanked him for his review and complained about the difficulties that foreigners have with the beautiful but complex English language. In his reply, Andy showed one of his most outstanding qualities, his generosity. He quickly offered to proof-read my articles and has done so ever since. Luckily, I am not a very prolific writer. Over the past six years we have developed mutual trust and built a friendship. We wrote an article together for "Watts Up With That?" and I hope we can write more, or even a book, once I get my first academic book on climate finished and published.

Andy is a much better writer than I am. His petrophysical training and experience make it easy for him to understand the complex physical intricacies of the greenhouse effect or atmospheric phenomena. But what really distinguishes him is his rare ability to explain it in easily understandable terms to a wide audience, something that most scientists lack. He is the perfect person to write this book, that I have had the pleasure to read before writing these lines. Andy perfectly understands the arguments of Karoly and Happer and does a great job of explaining them to the uninitiated. Climate is one of the most complex subjects the human mind can research; I know a little about it and can say that what Andy does, explaining a high-level climate debate, is not easy.

This is an important book, because the society has been robbed of its right to a wide scientific debate on an important subject, even though they are paying for only one side of the research. This is a topic that will likely significantly affect their standard of living, the consequences are huge—honest debate is critically needed. We all know the skeptical side of the debate has been insulted, ridiculed, and attacked for years, yet a few courageous scientists, like William Happer, still persist with nothing to gain personally, except like Galileo they defend the truth against the dogma. Andy's approach to writing about the debate is so effective that it surprised me. Instead of reproducing the debate and then extending the arguments with lengthy explanations that would have tired most readers and made the book very long, he directs the interested reader to a reproduction of the full debate in pdf format at his website (andymaypetrophysicist.com). He then devotes the book to explain the scientific arguments of each side and the supporting evidence, making it an easier and more educational read.

As I know the arguments of this debate well, I can appreciate the good job the book does in presenting them in a fair and comprehensive way. No argument is ignored, summarily dismissed, or overestimated. Each one is analyzed and the evidence supporting them is presented. As a scientist I am only interested in the truth, and I have never had any problem with anthropogenic CO_2 being the cause for all the recent warming. It is just that, like William Happer, I find that it is not what the evidence says. I appreciate that Andy, from his personal position on the debate, has been very fair to both sides and I expect readers will appreciate it too. I am afraid such a level of fairness cannot be hoped for or expected from the other side of the debate.

For someone like me who has studied climate change at a scientific level for about six years, reading a book about two good scientists debating about climate change appeared only mildly interesting. Instead, the reading has been very pleasant, like listening to a musician playing a piece we know by heart and appreciating the skill and passion he puts into it. It is an exercise in how complex issues can be explained to those that have the right to know because they are the ones paying for the scientific research. This is the third book of Andy's I have read. All of them are very well written and each addresses a different aspect of the climate change socio-political-scientific phenomenon. I've liked them all, but as a scientist this is the one that most interested me.

Javier Vinós, PhD.
Molecular Biology researcher
Climate researcher

Madrid, SPAIN
January 21, 2022

Abbreviations used in this book

°C/2xCO$_2$:	Units of climate sensitivity, ECS, and TCR. It means degrees C per doubling of the atmospheric CO_2 concentration.
°C:	Degrees Celsius, 9/5 times degrees Fahrenheit (°F).
ACRIM:	The Active Cavity Radiometer Irradiance Monitor satellite instrument. It measures TSI. The acronym also means the related TSI composite record by Willson and Scafetta.
AGW:	Anthropogenic Global Warming.
AMO:	Atlantic Multi-decadal Oscillation.
AR4:	The fourth IPCC scientific assessment report of climate change (IPCC, 2007b).
AR5:	The fifth IPCC scientific assessment report of climate change (IPCC, 2013).
AR6:	The sixth IPCC scientific assessment report of climate change (IPCC, 2021).
ARGO:	Wandering robotic floats that measure ocean temperatures from the surface to 2000 meters. Argo is named after the Greek mythological ship that Jason sailed around the Greek islands.
CAGW:	Catastrophic Anthropogenic Global Warming.
CERES:	Clouds and the Earth's Radiant Energy System satellite instrument.
CMIP:	The WCRP Coupled Model Intercomparison Project. Three versions are mentioned in this book, version 3 is used in AR4, version 5 in AR5, and version 6 in AR6. The WCRP is independent of the IPCC.
CRE:	The net Cloud Radiative Effect as measured by the CERES satellites. When positive, more energy is retained by Earth than emitted, a warming effect.
CS:	Climate Sensitivity to CO_2.
CSIRO:	Australian Government's Commonwealth Scientific and Industrial Research Organization.
EBAF:	Energy Balanced and Filled, processed CERES data.
ECS:	Equilibrium Climate Sensitivity, this is the ultimate warming after an instantaneous doubling of the CO_2 concentration.
ENSO	Stands for El Niño-Southern Oscillation, or the El Niños and La Niñas.
GHE:	Greenhouse Effect.

GHG: Greenhouse Gas.

GMST: Global Mean Surface Temperature, this is constructed from LSAT and SST.

GSAT: Global Surface Air Temperature at two meters. This is the air temperature at two meters over both land and ocean.

HadAT2:   Hadley Centre Atmospheric Temperature dataset 2 weather reanalysis.

HadCRUT: The combination of the sea surface temperature records from the Hadley Centre of the UK Met Office (HadSST) and the land surface records from the Climatic Research Unit of the University of East Anglia. Two version are used in this book, version 4 and version 5.

hPa: Hectopascals. This unit of pressure is equal to millibars (mbar) but is often preferred because it is an SI (International Standard) unit.

IPCC: The U.N. Intergovernmental Panel on Climate Change.

IR: Infrared radiation.

JAMSTEC: Japanese Agency for Marine-Earth Science and Technology.

JASON: A small independent group of elite scientists that advise the U.S. government on science and technology. Dr. Happer has been a part of JASON since the 1970s.

km: Kilometers, about 0.62 miles.

kyr: Thousand years ago.

LIA: Little Ice Age (~1500 to 1850AD)

LSAT: The average land air temperature at 2 meters (the surface).

Ma: Millions of years ago.

mbar: Millibars of pressure, one-thousandth of the atmospheric pressure at sea level.

MIMOC: NOAA Monthly Isopycnal/Mixed Layer Ocean Climatology

MWP: Medieval Warm Period (~900 to 1200AD)

Myr: Millions of years.

NAIP: North Atlantic Igneous Province.

NCEP: National Centers for Environmental Prediction weather reanalysis.

NGRIP: North Greenland Ice Core Project.

NH: Northern Hemisphere.

OLR: Outgoing longwave-infrared radiation from Earth.

PDO: Pacific Decadal Oscillation.

PETM: The Paleocene-Eocene Thermal Maximum, a warm period about 55.6 million years ago.

PMOD:	The *Physikalisch-Meteorologisches Observatorium Davos* TSI reconstruction.
ppm:	parts per million.
RAOBCORE:	Radiosonde Observation Correction using Reanalysis dataset.
RICH:	Radiosonde Innovation Composite Homogenization model ensemble. RAOBCORE and RICH are based mostly on weather balloons.
RSS:	Remote Sensing Systems' satellite estimates of atmospheric temperature. The work is headed by Carl Mears.
SAR:	The second IPCC scientific assessment report of climate change (IPCC, 1996).
SB:	Stefan-Boltzmann equation or constant.
SH:	Southern Hemisphere.
SILSO:	Sunspot Index and Long-term Solar Observations.
SST:	Sea Surface Temperature, nominally at a depth of 20 cm.
SW:	Incoming shortwave radiation from the Sun.
TAR:	The third IPCC scientific assessment report of climate change (IPCC, 2001)
TCR:	Transient Climate Response, this is the warming at the time of CO_2 doubling following a linear increase in CO_2 forcing over a period of 70 years (about 1% per year).
TOA:	Top Of the Atmosphere, generally defined as the orbit of the CERES satellites.
TPW:	Total Precipitable Water in the atmosphere, the normal units are mm or cm and independent of area. In this book we use Kg/m^2.
TSI:	Total Solar Irradiance striking the top of the atmosphere.
UAH:	University of Alabama-Huntsville, where John Christy and Roy Spencer estimate atmospheric temperature using satellite data.
UHI:	Urban Heat Island effect, that is cities are warmer than the surrounding countryside.
UNFCCC:	The United Nations Framework Conventions on Climate Change, it also refers to the U.N. secretariat for climate change.
$W/m^2/°C$:	Watts per meter squared per degree Celsius of warming. This is the feedback due to one degree C of warming.
W/m^2:	Watts per meter squared, the measurement units of radiation leaving Earth (OLR) and incoming (SW).

Preface

Are humans causing dangerous changes to Earth's climate by burning fossil fuels? Are the dangers, if any, consequential enough to cause us to stop using fossil fuels? What are the benefits of continuing to use fossil fuels? Are the benefits worth more than the dangers' cost? Are human-caused climate changes, if any, greater than normal natural climate variability? Is more CO_2 in the atmosphere a good thing or a bad thing? These are questions from one of the most consequential debates of the modern era.

February 15, 2016 was the beginning of an in-depth debate on man-made climate change between two well-known experts in the field, Dr. William Happer and Dr. David Karoly, hosted by James Barham and his team at TheBestSchools.org. Both have been heavily involved in atmospheric research since the 1980s. Happer believes that burning fossil fuels will have a minimal effect on climate but a large benefit to plant life and humanity. Karoly believes the opposite.

Dr. William Happer is the Cyrus Fogg Bracket Professor of physics, Emeritus, at Princeton University. He is an expert on the effects of radiation striking molecules in the atmosphere. This is a key—perhaps *the* key—element in this debate.

Dr. David Karoly recently retired as the Chief Research Scientist at the Australian Government's Commonwealth Scientific and Industrial Research Organization (CSIRO) Climate Science Centre; and is currently an honorary Professor in the School of Geography, Earth, and Atmospheric Sciences at the University of Melbourne. He is an expert in the planetwide effects of changes in Pacific Ocean surface temperatures. Another key element in the debate.

I first wrote about this debate in 2018 (May, 2018d), but felt it needed a more serious look. Debates on climate change between scientists as prominent and qualified as Happer and Karoly are exceedingly rare, which is a shame. Debate between experts is the best way to resolve contentious and complex disagreements. Millions of lives and livelihoods potentially hang in the balance. Climate is always changing, so the essence of the debate is whether humans are dangerously changing climate by burning fossil fuels and, if so, whether the best solution is for governments to curtail their use.

How certain is the conclusion by some scientists that burning fossil fuels will lead to a climate disaster? Only debates can ferret out their certainty or lack of it. Burning fossil fuels may cause some harm, but if we stop burning them, we will face certain harm. Which is worse? Debates educate the public, they are necessary.

Climate alarmists are a radical fringe of the so-called "consensus." They believe the science is settled and there is no need for further debate. In this book we try and distinguish between those who are simply concerned about man-made climate change, but want to learn more, and those who are so certain they want to shut down debate and end fossil fuel use immediately. We call the latter group the alarmists.

When skeptical people challenge the alarmist view that fossil fuels are causing a climate disaster, they are called "deniers" or other names and ridiculed. But, seriously, how certain is this radical consensus position? If they are correct, why can't they explain their position in clear, definitive language?

Einstein was certain of his theory of relativity, he could explain it, and eventually everyone understood it well enough to accept it. The theory of relativity is not proven, scientific theories are never proven, but we've seen enough confirming evidence that it is universally accepted. This is not true of human-caused climate change, thousands of well-qualified scientists, like Professor Happer, are skeptical.

There are almost 10 million people in the United States working in the oil, coal, and natural gas industries. According to the American Petroleum Institute, they comprise about 5.6% of total U.S. employment and they are paid above average wages. Are we to throw these people out of work because some people claim the "science is settled?"

We should always debate issues with such severe potential consequences. We are fortunate to have a detailed record of this one.

—Andy May
Author of *Climate Catastrophe! Science or Science Fiction?*
Blood and Honor: The People of Bleeding Kansas
Politics and Climate Change: A History

Chapter 1: The Great Debate

King Janaka, ruler of India in the seventh or eighth century BCE, was known as a great philosopher-king. He sponsored and sometimes participated in debates called "Vadavidya." They were considered the way to the essence of truth. Critical thinking, reasoning, and strict rules were the tools of Vadavidya. (Tripathi, 2016, pp. 2-3)

Introduction

Debates between prominent scientists on global warming or human-caused climate change are rare. Anthony Watts believes they are rare because the climate establishment nearly always loses them (Watts, 2018). On March 14, 2007, there was a famous debate held at the Asia Society and Museum in New York City, sponsored by the U.S. debating organization, Intelligence Squared, or IQ2US (IQ2US, 2007). It was three-on-three, moderated and hosted by Brian Lehrer. The climate skeptics, who were for the debate resolution "Global Warming is not a crisis," were Professor Richard Lindzen, Professor Philip Stott, and author Dr. Michael Crichton. Their opponents were NASA climate scientist Dr. Gavin Schmidt, Union of Concerned Scientists member Dr. Brenda Ekwurzel, and Scripp's Institute Professor Richard Somerville.

Prior to the debate, 30% of the audience were for the resolution and 57% were against. This was according to the official debate transcript (IQ2US, 2007), NPR (Malakoff, 2007b), and *Forbes* (Taylor, 2011). After the debate, a new poll of the audience showed that 46% were for the assertion that "Global Warming is not a crisis," an increase of 16%. Only 42% were against the assertion, a drop of 15%. The number of undecideds hardly changed, it was 13% before and 12% after (Lehrer, 2007). Thus, in this very high-profile debate, the climate skeptics won a huge victory. The establishment climate consensus have been reluctant to debate publicly ever since.

Oddly, the Intelligence Squared website, on December 28, 2020, reported completely different results that were clearly incorrect (IQ2US, 2007). The Internet Archives (Internet Archive, 2021) shows that the original web posting reflected the correct results, but the web site, as of August 26, 2021, reported that 43% were opposed to the resolution pre-debate and 89% were opposed post-debate, the opposite of the original results. The original results match the transcript and recordings of the debate. They are also the results reported by the debate moderator, *NPR,* and *Forbes.* Unfortunately, this sort of lying has become all too common in the public climate debate. This author and others notified them of their web error on the IQ2US web site, but the errors are still there as of this writing. More details are available in a web post (May, 2021). After 9 months, we must assume the incorrect reporting is intentional.

In 2010, the film director James Cameron challenged Marc Morano, Andrew Breitbart, and Ann McElhinney to a debate on climate change (Morano, 2010). The debate was to be held at the Aspen American Renewable Energy Day summit in Colorado and was widely publicized. After setting up the debate, Cameron changed the format multiple times, first the press would be allowed in, then they were to be excluded. Cameron would be accompanied by two climate scientists, often changing the participants. The three climate skeptics agreed to all the changes. At the last minute, when Morano was already in the air on his way to Aspen, Joe Romm, of Climate Progress urged Cameron to cancel. He reportedly told Cameron, "We always lose these debates." (Romm obviously agrees with Watts, perhaps they always lose because they are wrong?)

Since Morano arrived after the cancelation and had to pay for his trip, he was offered 90 minutes to make his case, even though Cameron was not present. He tried, but the moderator, Richard Greene, and the audience constantly interrupted him and would not let him finish his speech. Cameron backed out, thus forfeiting his position, Morano graciously tried to make his case to the audience but was ridiculed and shouted down—pretty much the way it has been ever since. If you do not have evidence on your side, all you can do is ridicule and shout down your opponent.

In a proper debate, appeals to authority, like "97% of scientists agree …" and name calling ("You're a denier") don't get you very far. You also can't call upon a compliant and ignorant news media to ridicule your opponent until he quits. You need facts, solid data, and documented examples. The climate consensus have none of this. As we will see in this informative debate between distinguished climate scientists Professor William Happer and Professor David Karoly, observations are mostly on one side—the side of the skeptics. The entire case of an impending human-caused climate disaster is based upon unfounded conjecture.

TheBestSchools

TheBestSchools is a company that helps students find the best college or university. They rank schools and provide prospective students with the information they need to choose the right school. February 2016, James Barham, of TheBestSchools, began what he called a "Focused Civil Dialogue" on global warming between Professors William Happer and David Karoly. It was a debate, conducted mostly by email, that involved an interview with each participant. The debaters then each wrote a statement on the topic, a detailed response to the opposing statement, and a final reply. All documents were placed on TheBestSchools.org web site.

I wrote previously about this Focused Civil Dialogue (May, 2018). However, when Dr. Martin Capages, my publisher at the time, encouraged me to write this more formal book on the debate, I found that some of the site's web pages had disappeared. Several illustrations and links in the remaining web pages had disappeared as well. Fortunately, I was able to recover the missing information using the "Wayback Machine" (Internet Archive, 2021). The nine original web pages were recovered, carefully checked, and turned into pdf documents. Except for reformatting, minor editing to correct typos, misspellings, and grammatical errors, they are as they were on the web. Editorial additions are in italics and square brackets and the bibliographies were completed. These documents are on the author's website, they are listed below and the internet URL for each of them is in the bibliography.

1. Introduction to the debate (TheBestSchools.org, 2021b)

2. William Happer's Interview (Happer, 2021a)

3. David Karoly's Interview (Karoly, 2021a)

4. William Happer's Major Statement (Happer, 2021b)

5. David Karoly's Major Statement (Karoly, 2021b)

6. William Happer's detailed response (Happer, 2021c)

7. Glenn Tamblyn's detailed response (Tamblyn, 2021a)

8. William Happer's final reply (Happer, 2021d)

9. Glenn Tamblyn's final reply (Tamblyn, 2021b)

All the documents were sent to James Barham, William Happer, and David Karoly in February 2021 for them to check for accuracy, if they wished. No problems with the documents were reported. I repeatedly tried to communicate with Glenn Tamblyn, but he never responded to my emails.

As explained in the "Introduction to the debate" document (item 1 in the list above), David Karoly dropped out of the debate in early 2017 after reading Happer's Major Statement. According to Barham, Karoly just stopped responding to emails and phone calls and offered no reason for dropping out.

Karoly responded to my introductory email and offered to send me the email to Barham giving his reasons for leaving the debate. Barham does not remember receiving it and does not have a copy, so I enthusiastically accepted Karoly's offer. He did not send it, I followed up a week later, but he was not forthcoming. So, the reason or reasons for Karoly's withdrawal, are known only to him.

Karoly is well trained in physics, and possibly he backed out due to Happer's Major Statement. Happer's statement is devastating to the consensus position, but hard to understand. It is highly mathematical and requires a good education in physics. However, when understood, it is convincing. It is discussed in Chapter 16, but without most of the equations. If I were in Karoly's position, I might have backed out after reading it as well.

After Karoly backed out of the debate, James Barham made many attempts to reach him—both by phone and by email—but Karoly never responded. So, rather than end the debate after Karoly's withdrawal, TheBestSchools recruited Glenn Tamblyn, an engineer who blogs about climate science at a website named skepticalscience.com. The site name suggests it is skeptical of "catastrophic anthropogenic global warming" (CAGW), but the blog promotes and supports it, as does Glenn Tamblyn. A brief biographical sketch and a list of his blog posts are on their website (Tamblyn, 2021). Tamblyn received a degree in mechanical engineering from Melbourne University and worked in the solar energy and automotive industries. Tamblyn has been blogging since 2011 and is a co-author of the book, *Introducing Climate Science* (Mason, Painting, & Tamblyn, 2018).

This debate is a valuable reference because both principal debaters are well-known scientists with a deep understanding of the emerging field of climate science as well as the more established field of atmospheric physics. Glenn Tamblyn is not in the same league as Dr. Happer and Dr. Karoly, but he does have an excellent understanding of what we might call "mainstream climate science." He makes the arguments that we often hear from the popular press and politicians and brings in topics that may have been ignored had the debate only been between Karoly and Happer.

Tamblyn started over when he entered the debate and this forced Happer to argue two debates, rather than one. Tamblyn often did not fully understand what Happer had written, which led to some confusion, but Happer soldiered on and completed the task to which he agreed. Tamblyn added several important issues that would not have been debated had he not been included.

Both James Barham and Dr. Will Happer have seen the book and identified no problems with its content. So, I sent Dr. Karoly one last email asking if he would review it and offered him the opportunity to write a Foreward for the book. Since I had not heard from him for nearly a year, even though I had sent him two emails, I did not expect a reply. But thought I should offer him a final chance to comment. The book is about the importance of debate in science and having his rebuttal to the book right in front of it, fits right in with the book's thesis.

At first, he agreed, he asked for a month to review it and then another month to write the Foreword, I agreed to his terms and waited a month before contacting him again. He then replied that he needed a few more days to complete his review of the book, but promised to send his review by January 3, 2022, right after the New Year. He claimed to have identified "a number of factual errors" in the book but did not specify what they were. That was the last I heard from him, so he was given every opportunity to respond, but ghosted me, just as he had done to James Barham.

Book Structure

The documents we worked from are listed above. They are organized by person and debate task. This book is organized by the questions asked by James Barham and TheBestSchools. Following the questions, I provide the answers from the debaters who offered them. Typically, after the questions and answers, I provide additional background on the question. The background sections have a minor heading that begins with "Background:" These sections are written by me, not the debaters, and can be skipped if the reader just wants to read what the debaters have to say. The background sections contain more current and up to date references and fill in critical details and context left out by the debaters.

All statements and opinions from Karoly, Happer and Tamblyn are identified as such. We also identify quotes and ideas from many other scientists in the text. Assume that any opinions or information not attributed to someone else, are from me. Because Barham, Karoly, Tamblyn, and Happer are addressing each other, they often speak at a high scientific level, so some of the concepts debated required further explanation and the background sections are intended to provide that. The background sections are underlined in the index to make them easier to find.

We've already provided the information we have on Glenn Tamblyn's background. Next, we will discuss the backgrounds of Dr. Happer and Dr. Karoly plus their introductory remarks.

William Happer

William Happer is the Cyrus Fogg Brackett Professor of Physics, Emeritus of Princeton University. He has specialized in the interactions of radiation with matter and is a member of JASON, a small group of scientists who advise the U.S. government on science and technology, especially as it relates to national security. Happer was also Director of the U.S. Department of Energy's Office of Science from 1991 to 1993.

The details of Happer's life story presented here are from his interview (Happer, 2021a) and an email discussion with him. Happer was born in Vellore, India on 27 July 1939. He was the son of Dr. William Happer, a Scottish medical officer in the Indian Army, and Dr. Gladys Morgan Happer, a medical missionary for the Lutheran Church of North Carolina, USA. World War II started a month after his birth when Hitler and Stalin invaded Poland.

At that time, the Japanese were invading China and the fighting was fierce. The Chinese had won their first battles against the Japanese in 1939 and were preparing their first large counterattack. Although the counterattack failed, it did slow down the Japanese.

Even so, the progress of the Japanese army in China led to persistent rumors that they would attack and conquer India. Happer's mother became pregnant with his brother Ian and his father was concerned for his family's safety. So, early in 1941, he put Gladys and one-year-old Happer on a ship bound for America.

The ship, carrying the pregnant Gladys, who was caring for her infant son, had to make it alone through dangerous submarine-infested waters to the United States. It carefully rounded South Africa and sailed up the eastern coast of South America to avoid German submarines and eventually made it safely to North Carolina. The family then made their way to Gladys's parents in Salisbury, North Carolina, north of Charlotte.

In the meantime, Happer's father, William Happer, Sr., was sent to Iraq, which sided with Hitler during the war. After that posting, he was sent to Egypt to help fight Rommel's army.

Happer's uncle, Karl Ziegler Morgan, was one of the first physicists to join the Manhattan Project in Oak Ridge, Tennessee. He was put in charge of protecting the workers from radiation. Later, Karl Morgan would be known as the "Father of Health Physics." He was knowledgeable about nuclear physics but knew little about medicine, so he asked his sister, Happer's mother Gladys, to join him at Oak Ridge in the fall of 1944 (Howes & Herzenberg, 2003, p. 125). Gladys was the first physician to arrive at the Oak Ridge facility.

Karl Morgan described his sister Gladys as the "smartest person I have ever known." He said she had a remarkable memory and "most of the brains in the family." She graduated at the top of her class at the University of North Carolina with an MS in organic chemistry. She then went on to take her medical degree at the Woman's Medical College of Pennsylvania in Philadelphia. The University of North Carolina did not want to recommend her to a major medical school because she was a woman.

After graduating with her medical degree, she did some postdoctoral work in Paris and Geneva and enjoyed her time there. But the Lutheran Church had paid for her medical school and sent her to India as a medical missionary where she met her husband .

Five-year-old Happer went with his mother to Oak Ridge and had a series of baby-sitters while his mother worked in "X-10," the facility that would later be known as the Oak Ridge National Laboratory. By all accounts, Gladys was a wonderful doctor. Her patients and fellow employees said she was the best doctor in the facility. Happer's Uncle Karl became his father figure at the time, since young Happer didn't know if his real father was alive or dead. It was at this time young Happer decided to become a nuclear physicist, like his uncle.

Fortunately, the elder William Happer did survive the war and brought the family back to India. Young Happer attended the International School at Kodaikanal in the mountains of southern India. The family remained there until 1948 when India gained its independence. Then they returned to Scotland and stayed with the elder Happer's mother in Edinburgh, both Ian and young William attended the James Gillespies School there.

Scotland was still recovering from the war and with all the returning veterans and colonials, it was hard for the elder Happer to find a good job. Shortly after arriving in Scotland, the family was blessed with the birth of Elizabeth, William's younger sister, on August 23, 1948. Since the elder William Happer could not find a good job and Gladys missed her home, the family set sail for America in 1950.

The Happers had been financially broken by the war and were nearly penniless. They accepted help from Gladys's family in the United States. Happer's father, then 50 years old and a member of the prestigious Royal Society of Physicians and Surgeons, was not qualified to work as a medical doctor in the United States, so he had to study for several months to take the U.S. medical exams. He succeeded, receiving the highest score of his group.

The senior William Happer, upon receiving his license to practice, accepted a job to head the public health department in Caldwell County, North Carolina. It was there, at the edge of the famous Blue Ridge Mountains that young William went to high school.

Young William also worked as an assistant beekeeper to earn money for college. He won a scholarship to study physics at the University of North Carolina. After graduating, his Uncle Karl encouraged him to go to Princeton for graduate school, where he earned his PhD in nuclear physics in 1964.

His thesis and most of his subsequent work was on the interaction of radiation and matter, a narrow field of atmospheric (or nuclear) physics that is key to understanding greenhouse gases and how they might cause atmospheric warming. Certain wavelengths of incoming solar radiation and radiation emitted by Earth's surface can be captured by CO_2, water vapor, and other greenhouse gases and later re-emitted. The small amount of time between the energy capture and later re-emission, causes the excited molecules to warm the air around them (Pierrehumbert, 2011).

After receiving his PhD, Happer took a postdoctoral research position at Columbia University and studied optical pumping, a process that uses light to raise electrons in an atom to a higher energy level. This technology is used in laser construction. While at Columbia, Happer met Barbara Baker, a nurse at Columbia Presbyterian Hospital. They fell in love and were subsequently married in New York and had two children, Jim and Gladys.

Happer was soon promoted to Assistant Professor at Columbia. The young family settled into an apartment in New York and were doing well. Then, on April 24, 1972, a group of 40 Columbia faculty members and a few students invaded and took over Pupin Hall, the Physics building on the

Columbia campus. They imprisoned several professors involved with the JASON Division of the Institute for Defense Analysis (IDA). Happer was not yet a member of JASON but was gathered up with the rest. While imprisoned, the prisoners were "charged" with participating in the Vietnam War and harangued about it.

The first night of Happer's captivity, Alan Ginsburg arrived at Pupin Hall and, accompanied by an accordion, recited a poem about "Academic War Criminals." Happer felt the performance somewhat defused the tension in the hall.

On the first day of the occupation, Columbia took no action. However, on the second, they sent in the police causing a near riot, so the police withdrew. After the withdrawal, the protestors chained all the doors and prepared for a long siege.

That night, Happer managed to escape from Pupin Hall and tried to walk home to his wife and young children. As he walked across campus, a gang of teenagers from Harlem, grabbed him and dragged him into an area where a wild-eyed white guy was giving a speech about "white oppression." The teenagers took Happer to the speaker and yelled "we got one, boss, we got one!" This interrupted the speaker, at an intended climax of his speech, and he was visibly annoyed. He asked Happer "who are you?" Happer responded "I am Dr. William Happer, an Assistant Professor of Physics." The speaker was unimpressed, and with a sneer, he told the teenagers "Let him go."

This was a lucky break, as he could have been beaten or worse. But he was young, and more than a little upset being seen as too small a fish to bother with.

The next day, April 27, a large group of conservative students and faculty burst into Pupin Hall with fists flying, beating, and dragging the protestors out of the building. Afterwards, they proudly announced that Pupin was cleared. The protestors had wanted the Pupin Hall professors to resign from JASON and the IDA, none did. The whole experience was one reason why Happer later joined JASON and conducted confidential research for the Pentagon and U.S Government for many years.

Happer continued working at Columbia and eventually achieved the rank of full professor. In 1980 he moved to Princeton University where he became the Cyrus Fogg Bracket Professor of Physics.

His work on the interaction of energy and matter in the atmosphere led to a key insight allowing development of the "Sodium Guide Star" technology for which he is now famous. He first presented this idea at a secret JASON conference in 1982. The work was done for the Defense Department and remained classified for a long time. When it was finally declassified, in the 1990s, Happer received some well-deserved public recognition for his work.

The technology is used to guide missiles to their target and to improve the quality of images taken with large telescopes. It removes or greatly reduces the effect of atmospheric turbulence on light. Happer also used his knowledge of energy and atmospheric interactions to study climate change. He was a co-author of the 1981 book *The Long-term Impacts of Increasing Atmospheric Carbon Dioxide Levels*, edited by Gordon MacDonald (MacDonald, 1981).

Happer was made Chair of the JASON Steering Committee in 1987 and appointed Director of the Office of Energy Research at the U.S. Department of Energy by President George H. W. Bush in 1990. He was carried over into the Clinton administration until Vice-President Al Gore abruptly ordered Hazel O'Leary to fire him. She offered Happer another position, since he was well thought of, but Happer decided to go back to Princeton University.

After returning to Princeton, he was horrified at the illiterate news media and politician's distortions about CO_2 and climate change. The stories about climate doom increased dramatically after Al Gore took office and began hiring climate activists and firing the more realistic scientists like Happer. William's wife, Barbara, encouraged him to speak out and he did.

It did not take long for Gore and his "climate establishment" to attack Happer. Their usual attacks were based upon the fact that he was not a member of their self-appointed group of "climate scientists." Traditionally, climate was studied by Earth scientists, especially geologists, meteorologists, atmospheric physicists, and geophysicists. However, once people began writing climate computer models, the small group of modelers began calling themselves climate scientists and tried to exclude actual scientists from the club.

Helping Gore and the so-called climate scientists in their efforts to demonize skeptics were several non-profit environmental groups like Greenpeace. Greenpeace, the Union of Concerned Scientists, and other radical left wing environmental organizations have made a successful business of ridiculing, harassing, and even extorting money from businesses and individuals they disagree with. This demonization is explained in Chapter 3 of our book, *Politics and Climate Change: A History* (May, 2020c). A compliant news media has helped their efforts by publishing their press releases without checking the facts, as explained in Chapter 2 of the same book. In 2015, there were many attacks on mainstream scientists, Happer was not the only one.

Happer had a long history of promoting the benefits of additional CO_2 in the atmosphere but had no intention of peddling the views of someone else, only his own. The *New York Times* (Schwartz, 2015) claimed Happer was offered money by a Greenpeace operative to write a paper on the benefits of CO_2, and he accepted it. It is clear in the email chain (Happer, 2015) that he did not, he simply said that if any fee were offered, it should go directly to the CO_2 Coalition, a 501(c)(3) tax-exempt educational organization. Happer has never taken any money from anyone for his CO_2 research, other than occasional travel expenses.

Happer on the basics

Happer on the basics

TheBestSchools.org asked Happer to comment on the following:

"The IPCC's official position may be summarized as making four claims: global warming is a well-established fact; it is anthropogenic; it is a major problem for humanity; and concerted global governmental action is required to combat it.

"Critics claim that the computer models upon which the IPCC's official position is based are unreliable ... [since] the atmosphere is a [turbulent] fluid-dynamic system which makes its long-term behavior very hard to predict ...

"Supporters of the ... 'consensus' position ... would argue that such considerations are basically irrelevant, given the simplicity of the physics of the 'greenhouse effect.' (Happer, 2021a, pp. 16-17).

TheBestSchools then explains that the consensus tells us that CO_2 has been increasing and CO_2 is a greenhouse gas. Therefore, Earth's atmosphere must warm. Happer is comfortable with the consensus position that CO_2 has been increasing in the atmosphere mostly because of humans burning fossil fuels and he agrees that CO_2 is a greenhouse gas (GHG) that will warm Earth's atmosphere at the surface. Happer then points out that the magnitude of surface atmospheric warming caused by increasing CO_2 concentration is unknown.

He suspects that the ultimate warming, due to instantaneously doubling the CO_2 concentration (a quantity known as equilibrium climate sensitivity, or "ECS"), will be about one degree Celsius per doubling of CO_2, which is close to the direct warming expected from that CO_2 concentration absent feedbacks. ECS is the temperature change over the long-term, after the climate system has reached equilibrium with the new CO_2 concentration, this can take hundreds of years.

Happer and the consensus agree that the effect of CO_2 on atmospheric temperature varies widely with altitude and location, additional CO_2 may cool the stratosphere, especially above 20 km. Karoly believes that the direct warming, at the surface, from additional CO_2, will be increased by positive feedback. The idea is that the higher temperature, caused directly by CO_2, will increase the amount of water vapor—a more powerful GHG than CO_2—at the Earth's surface. This happens only in the lower atmosphere since water vapor condenses out at high altitudes. When it condenses it releases thermal energy, or "heat," that causes warming.

Happer explains that as time goes on there is less observational support for this positive feedback. In the lower atmosphere, roughly half the thermal energy is carried from the surface as latent heat in the form of evaporated water. This is the natural process of convection. However, while it does increase the concentration of water vapor near the surface, it also adapts to higher temperatures by forming clouds and haze, increasing Earth's reflectivity or albedo. The increased albedo allows less of the Sun's energy to strike the surface and warm it.

Happer and most other scientists believe that higher surface temperatures in the tropics, where most water vapor resides, increases precipitation efficiency. It rains more in the tropics than in the higher latitudes, improving the efficiency of energy transfer from the surface to higher altitudes. The increasing precipitation also increases the velocity of the natural water cycle.

It is generally recognized that clouds form high in the atmosphere where it is cool enough for water vapor to condense into water droplets and ice. As the water vapor changes to liquid and solid form, a cloud forms and the latent heat gained by the water vapor as it evaporates is released. Some of the latent heat escapes to outer space, some warms the surrounding air, and the remainder is carried to cooler areas by wind.

Clouds reflect sunlight during the day and warm the night air by absorbing and re-emitting thermal energy from Earth's surface. This moderates local temperature change.

Happer believes that both global warming and global cooling are well-established facts, and that the world is always warming or cooling. He disagrees with the IPCC[1] and believes most of the recent warming is natural, but acknowledges that some is probably due to humans, either through building cities and roads, or due to CO_2 emissions. The correlation between CO_2 concentration and surface temperature has been poor, according to Happer, at all timescales.

Happer, contrary to what the IPCC asserts, does not believe that the additional CO_2 emitted by humanity is a problem. He believes the additional CO_2 will boost agricultural productivity, make Earth greener, and benefit mankind. On average, the world will warm, but little of this warming will occur in the tropics and in the daytime, the warming will occur mostly in the polar regions and at night. These are the places and times most beneficial to humans.

Happer believes that government intervention is unnecessary, dangerous, and harmful to humanity. So far government actions have only driven up the price of electricity, disproportionately harming the poorest among us, and blighted the landscape with windmills and solar farms. Government interference in the scientific inquiry into climate change has corrupted the scientists involved and flooded them with money so they will produce politically correct results. According to Happer, "warming is not the problem, government action is the problem." He also believes "government actions have corrupted science."

---

[1] The United Nations Intergovernmental Panel on Climate Change

David Karoly

Dr. David Karoly advocates for the view that human-caused global warming or climate change is dangerous. Karoly is a meteorologist, an expert in climate change, and ENSO.[2] He has been involved heavily in several IPCC reports, both as an author and lead author. At the time of the debate, Karoly was the Professor of Atmospheric Science in the School of Earth Sciences at the University of Melbourne, Australia. After briefly leaving the University to become the Chief Research Scientist at CSIRO, he returned to the University of Melbourne as an honorary Professor.

The following brief synopsis of his early life is from his interview with TheBestSchools.org prior to the debate (Karoly, 2021a). Dr. Karoly was born in Sydney Australia in 1955 and lived there for the first eight years of his life. His father was an electrical engineer and worked as a computer programmer and engineer in the early days of computing. His father emigrated from Hungary to Australia in 1939, with his parents. The timing was good, as World War II started in eastern Europe shortly after he left.

Karoly's father finished high school in Melbourne and then went to the University of Melbourne and graduated with a degree in electrical engineering. Both of Karoly's paternal grandparents had university degrees from the University of Budapest.

Karoly's mother grew up in Melbourne with her Australian parents. She worked at a bank after graduating from high school and met David Karoly's father at a tennis club. After they married, they moved to Adelaide, then later to Sydney. In Sydney, Karoly's father worked on one of the first computers in Australia.

Karoly started elementary school in Sydney, but the family moved to Mountain View, California for a while when he was eight years old. They later returned to Australia and Karoly entered high school in Melbourne in 1967. Karoly describes himself as a nerd with a slide rule. Math and science came easily for him, and he settled on a scientific career while in high school.

He graduated from high school with good grades and received a scholarship to Monash University, a new school in Melbourne. He finished his general undergraduate degree in three years and was invited to do an Honours year in applied mathematics. His Honours project was in atmospheric and ocean fluid dynamics. It was during that year he decided to specialize in geophysical fluid dynamics and study the atmosphere.

He won a scholarship from Shell Australia to study in England. He chose to go to the University of Reading and entered their Meteorology Department. Karoly worked with Brian Hoskins, his thesis advisor, on how El Niño and La Niña affect the weather many thousands of miles away. The research was

---

[2] ENSO stands for El Niño-Southern Oscillation, this is the system that produces La Niñas and El Niños

successful and the original paper (Hoskins & Karoly, 1981) has been cited over 2,600 times.

He finished his PhD in 1980 and returned to Australia to a postdoctoral fellowship in the Australian Numerical Meteorology Research Centre, a joint center run by the Bureau of Meteorology and CSIRO (Commonwealth Scientific and Industrial Research Organization) in Melbourne.

David Karoly's grandfather was Alexander Karoly who worked for the Bureau of Meteorology in Victoria, Australia. Alexander had written some papers on drought and extreme rainfall. Unfortunately, David did not become aware of this until after his death. He wished he had known beforehand, so that he could have discussed his grandfather's work with him before he passed.

Karoly on the basics

Prior to the debate, **TheBestSchools.org** ask Karoly to comment on the following:

"You have stated that you started out as a global warming skeptic yourself. What changed your mind?

"… ought consensus scientists … [have] such a high degree of confidence in their computer models?" (Karoly, 2021a, pp. 10-11)

In the 1980s, Karoly was skeptical that GHGs were contributing to global warming. But, at the time, he was also researching Southern Hemisphere atmospheric temperature patterns and noticed that, while surface temperatures were warming, stratospheric temperatures were cooling. In his view, this pattern was not consistent with natural warming, but more consistent with warming due to increasing atmospheric greenhouse gas concentrations. After completing his research, Karoly concluded recent warming was due mostly to human activities and increasing atmospheric GHGs.

Karoly's logic is clear. If CO_2 and other GHGs increase, they warm the lower atmosphere because they capture incoming and outgoing infrared radiation, become excited, and warm the air around them. They also emit infrared radiation, but at the surface, little of that radiation can make it to outer space because the lowermost atmosphere is too dense and contains abundant water vapor over most of Earth's surface. Surface water vapor and CO_2 block most outgoing infrared radiation. However, in the stratosphere, the air is very thin, there is very little water vapor, and emissions from CO_2 cool the air. Thus, adding CO_2 to the stratosphere, increases emissions to outer space, resulting in a net cooling effect. Karoly calls this simultaneous warming of the troposphere and cooling of the stratosphere the "fingerprint" of human-caused climate change.[3] He published two papers[4] in the 1980s on this topic.

---

[3] (Karoly, David Karoly Interview, 2021a, pp. 11-12)
[4] (Karoly, 1989) and (Karoly, 1987)

Others were also pursuing this research, and as Karoly mentioned in his interview (Karoly, 2021a), the Second Assessment Report of the IPCC (SAR) used his and others' research to conclude that:

> "The balance of evidence suggests a discernable human influence on climate." (IPCC, 1996, p. 4).

This conclusion, written in bold type on page 4 of the Summary for Policymakers in SAR, came from the following statement in Chapter 8:

> "The body of statistical evidence ... points towards a discernable human influence on global climate." (IPCC, 1996, p. 439).

Background: The SAR atmospheric fingerprint controversy

The statement quoted above is from the final version of SAR, published in November 1995 and is different from what the original authors of SAR Chapter 8 agreed to in their final draft of July 1995. The author's final draft concluded:

> "...we have no yardstick against which to measure the manmade [climate] effect. If long-range natural variability cannot be established..." (Lewin, 2017, p. 277).

Thus, the scientists who wrote and agreed to the final draft of Chapter 8, concluded they *cannot* compute the human impact on climate because the range of natural climate variability is unknown. David Karoly was one of the original authors of that chapter.

The Co-Chairs of the volume, John Houghton and Gylvan Filho, and Benjamin Santer, a lead author of Chapter 8, replaced the statement above, from the final July draft, with the prior following a furious late-night debate among the IPCC leaders. That argument took place in Madrid, on 29 November 1995 (May, 2020c, pp. 280-282).

It was very controversial. The new statement—saying the opposite of what the chapter authors had agreed to—was inserted by the political editors of the publication at the insistence of politicians without consulting with the authors: the scientists who did the work. The most vocal lead author, Benjamin Santer, had agreed to the original draft, in July, but changed his mind by November. Even after changing the wording at the insistence of the politicians, including some in the United States, he published a paper in *Nature* (Santer B. , et al., 1996a) that agreed with the original text (May, 2020c, p. 232). Houghton admitted to being pressured by politicians to change Chapter 8. A representative of the U.S. government commented that there were:

> "Several inconsistencies and stated that it is essential that the chapters not be finalized prior to the completion of the discussions at the IPCC Plenary [meeting of senior IPCC authors and political representatives of the participating countries] in Madrid [the November meeting], and that the chapter authors be prevailed upon to modify their text in an appropriate manner following the discussion in Madrid." (Houghton, 1996)

Thus, the U.S government, and perhaps other countries, were pressuring Houghton to overrule the scientists and "prevail" upon them to change the science to match the political conclusions. There is no other way to read that. Houghton adds in a letter to *Nature* in 1996 that "proposals for modification to the draft Chapter 8 were made by both scientists and government delegates." The IPCC reports are presented as compilations of peer-reviewed scientific work. The reports are either scientific or political, they cannot be both.

When the political changes to scientific work became public, it caused an uproar. The head of the National Academy of Sciences, Frederick Seitz, spoke out forcefully and wrote:

> "In my more than 60 years as a member of the American scientific community, including service as president of both the National Academy of Sciences and the American Physical Society, I have never witnessed a more disturbing corruption of the peer-review process than the events that led to this IPCC report [SAR]" (Seitz, 1996)

Later work by Patrick Michaels and Paul Knappenberger (Michaels & Knappenberger, 1996) and Gerd Weber (Daly J. L., 1997) showed there were alternate explanations for Karoly's "atmospheric fingerprint" and that the data used to establish it had been cherry-picked. This ugly fingerprint controversy is explained in more detail in my earlier book (May, 2020c, pp. 230-235). Here we only need to point out that, while Karoly's idea is logical and compelling, it is not quantitative or definitive, it cannot even be definitively attributed to human environmental influence.

The fingerprint idea was also not supported as proof of anthropogenic influence by most of the authors of Chapter 8. They did not think that natural variability was understood well enough to definitively say that humans are influencing the climate.

The IPCC politicians prevailing upon the scientists to change their conclusions did great harm to the reputations of the IPCC and John Houghton personally. This event, more than any other in the history of the IPCC, caused scientists around the world to lose trust in the organization.

The controversy over the fingerprint and what it means or does not mean was widely publicized and discussed long before the Karoly-Happer debate started. It is quite surprising to see Dr. Karoly mentioning it in his opening remarks as his primary proof that humans are causing global warming.

Karoly on model accuracy

TheBestSchools next comment was about the accuracy of the models. Karoly acknowledges that the atmosphere is a complex, non-linear, fluid-dynamical system that cannot be predicted accurately. But he believes that with proper models bounds can be set, and these can be used to project average conditions.

Karoly believes that to model the climate system, the fluid dynamics of the atmosphere and ocean need to be described using the equations of motion. Next the energy losses and gains due to radiation, clouds, rain, snow, ice, and their interactions must be described. The changes in the energy losses and gains in the system determine the changes in the climate, and some of these are predictable if the forcings are known, even though the system is very complex. He concludes that the day-to-day and even the year-to-year variations of this chaotic system are not predictable.

The limits on the predictability of the long-term behavior of the climate system depend on the difference between forcing (like from changes in greenhouse gas concentrations) and the internal variability of the system at decadal timescales. The internal variability is much smaller at global scales than at a single location or a small region, smaller at decadal timescales than for a single year, and smaller for some variables, like temperature, than for others, like rainfall.

Scientists have confidence in their models because they test them and evaluate their performance using observations. This has been done using simulations of the climate of the 20th century and comparisons with observed climate variations. It is important to understand, when comparing climate models with observations, that both represent chaotic systems, and we shouldn't expect the precise state of the modeled atmosphere to agree with the observations at any one time. However, their average behavior should agree within the bounds associated with internal variability of the chaotic system.

Thus, Karoly believes that climate computer models are reliable because of how well they match average global observations. This opinion is controversial and will be discussed in more detail later in the book.

James Barham and the Debate

We are extremely fortunate to have this detailed record of a debate between two such prominent atmospheric physicists. The debate was mostly by email and moderated by James Barham. It began February 15, 2016. As mentioned above, Dr. Karoly backed out on January 20, 2017, in the middle of the debate, after Happer's Major Statement. The responses to Happer's statement and interview were written by Glenn Tamblyn.

David Karoly's major statement on climate change has been permanently deleted from TheBestSchools.org web site, but I have the detailed notes I took when I read it in 2018. I was also able to recover the text from the Wayback Machine and used the figure captions and bibliography to restore all the figures. The restored document is available on our website (Karoly, 2021b).

The debate is fascinating, but the material provided by TheBestSchools is long, repetitive, and poorly organized. Here we summarize the debate by the questions TheBestSchools asked. For each question, we present the arguments from Happer, Karoly and Tamblyn using their words whenever possible. The wording of the questions by TheBestSchools.org, sometimes slightly paraphrased, will also be provided.

The views of both scientists, and Tamblyn, are given for all questions they attempted to answer.

Happer and Karoly are familiar with the same data but use it to draw quite different conclusions. The different conclusions highlight the data's ambiguity, and a debate is the best way to learn how the differences came to be. We illuminate the differences by placing their arguments side-by-side.

Debates on climate change between prominent scientists are rare, so enjoy this one. In the next chapter we ask the debaters if recent global warming is unusual.

Chapter 2: Is Recent Warming Unusual?

The IPCC: "It is accepted that global-mean temperatures have increased over the past 100 years and are now warmer than at any time in the period of instrumental record." (IPCC, 1990, p. 254).

TheBestSchools:

"There are a number of inherent difficulties in obtaining reliable data about the past behavior of the atmosphere. Perhaps the most important of these is the weakness of the 'signal' (the secular change in temperature [is] measured in tenths of a degree Celsius), which is some three orders of magnitude smaller than the range of the geographic temperature variation, not to mention the normal diurnal and annual temperature variability, over the surface of our planet.

"Another issue is the relative difficulty in measuring surface temperatures over the ocean in comparison with the land. Yet another is the sparseness and non-random distribution of the land-based measuring stations where the majority of the data are gathered. For example, whether bias may have crept into the data due to the closing of weather stations in Siberia after the fall of the Soviet Union is one much-discussed problem. Bias due to the skewed placement of stations near urban areas (which are warmer than the surrounding countryside) is another. And so on.

"In light of all these difficulties, once again, ought we to place so much confidence in our models, which after all are only as good as the empirical data, we feed into them?" (Karoly, 2021a, p. 13).

While Happer and Karoly start from different places, they have many areas of common ground. They agree that climate changes, that the world has become warmer over the past 120 years, and that adding CO_2 to the atmosphere will cause some warming. They also fully agree that the CO_2 concentration in our atmosphere is increasing by about 2.5 ppm/year. About half of human emissions go into the atmosphere, the other half is absorbed by the ocean and biosphere. These facts are not in dispute.

Projected versus Actual Impacts

They differ on the *projected* impacts of potential future warming and additional CO_2. Happer thinks the impacts will be net beneficial and Karoly thinks they will be detrimental to humans and nature. They also differ on the magnitude of warming due to additional CO_2, sometimes called the "climate sensitivity," to doubling atmospheric CO_2. Happer thinks the climate sensitivity is low, perhaps one degree Celsius, while Karoly and the IPCC think

18

it is about three degrees Celsius. Karoly and Happer further disagree on the impact of additional CO_2 today, a topic we will discuss in later chapters.

We add context and background to the Happer versus Karoly and Tamblyn debate, and some commentary when we think it is needed. But, as much as possible, this book is about TheBestSchools (mainly James Barham), Happer, Karoly, and Tamblyn. Data and analysis from the fifth and sixth IPCC assessment reports (AR5 and AR6), other peer-reviewed references, books, and blog posts are added for clarity and context. In many cases we have added new references that were not available during the debate. The original debate documents, published by TheBestSchools.org, often reference journal articles, but the online bibliographies were incomplete. We located the missing references and added them to the original bibliographies in the restored original documents, listed in Chapter 1.

Recent global warming

Karoly compares the relatively accurate, high-resolution, modern, global-average-temperature rise of 0.9°C over the past 100 years to the sparse, low-resolution and poorly dated temperature-proxy records of the past 1,000 years and asserts that no 100-year temperature rise in the past millennium is as large as the current one. To quote Karoly:

> "There are a number of estimates of the hemispheric average temperature using different methods and different proxy data, not just the one shown [below, Figure 1] by Michael Mann and his collaborators. They all show that the period around 1000 AD was relatively warm and that the period around 1600 to 1800 was relatively cool, just as the Hockey Stick does. However, they all show that the increase in Northern Hemisphere average temperature over the twentieth century was larger than in any other century over the last millennium and that the last 30 years was likely warmer than any other 30-year period over the last 1000 years averaged over the whole Northern Hemisphere." From the Karoly Interview (Karoly, 2021a).

Karoly also emphasizes that temperature records are not "fed" into climate models. The model generated temperatures are compared to observations, but the observations are not model input.

Mann's Hockey Stick

The "Hockey Stick" graph Karoly refers to is originally from a paper published by Michael Mann, Raymond Bradley, and Malcom Hughes,[5] often abbreviated as MBH99. Figure 1 plots the same data but is from the IPCC Third Assessment Report (TAR) (IPCC, 2001, p. 29).

---

[5] (Mann, Bradley, & Hughes, 1999)

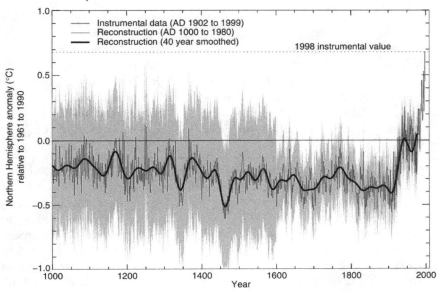

Figure 1. The "hockey stick" graph, from the third IPCC climate assessment report or "TAR" (IPCC, 2001, p. 29)

TheBestSchools asked Happer about the Hockey Stick:

"Critics say, among other things, that the Hockey Stick cannot possibly be right because it wipes out such well-known phenomena as the 'Medieval Warm Period' (c. 1001–1300 AD) and the 'Little Ice Age' (c. 1550–1850), as you have already alluded to above.

"Where do you stand on the Hockey Stick? Could you please explain to our readers what is right or wrong with it, and how (in your opinion) it came to have the significant role it has assumed in the discussion of global warming?" (Happer, 2021a, p. 35)

Happer replied:

"The hockey-stick temperature record was conspicuously absent from the latest IPCC report, which speaks volumes. My guess is that the hockey stick started out as an honest but mistaken paper, but one welcomed by the global-warming establishment. They had been embarrassed for years by the Medieval Warm Period, when Vikings farmed Greenland, and when emissions from fossil fuels were negligible. A.W. Montford's book, *The Hockey Stick Illusion* (Montford, 2010), is a pretty good summary of what happened." (Happer, 2021a, p. 35)

Background: The Hockey Stick

As Andrew Montford explains in the book Happer refers to, the statistical methods used to build the hockey stick were fatally flawed. Also, the tree ring data used to build the pre-1600 portion of the hockey stick were flawed and incompatible with 20th century data; a fact explained in a series of papers by Stephen McIntyre and Ross McKitrick (McIntyre & McKitrick, 2005). According to a legal statement by McKitrick, the events unfolded as follows:

"After publishing their 2003 *E&E* article [(McIntyre & McKitrick, 2003)] and reviewing Mann's unpublished responses to it, McIntyre and McKitrick [M&M] submitted an extended critique of the errors and misrepresentations in MBH98 [(Mann, Bradley, & Hughes, 1998)] to *Nature* magazine, which had published the first of the hockey stick papers. *Nature* solicited a response from Mann et al., and after examining it they ordered Mann et al. to publish a detailed correction and restatement of their methodology, which appeared in [(Mann, Bradley, & Hughes, 2004)]. M&M also extended their critique of Mann's statistical methodology and submitted it to *GRL [Geophysical Research Letters]*, which had published the 2nd hockey stick paper, and after peer review *GRL* published their study [(McIntyre & McKitrick, 2005)]. Mann et al. never submitted a response. A panel led by Professor Wegman later conducted an independent review of the mathematical and statistical issues and upheld the M&M critique [(Wegman, Scott, & Said, 2010)]. A panel of the National Academy of Sciences also conducted an examination of the whole issue of paleoclimate reconstructions and upheld all the technical criticisms M&M made of Mann's work, going so far as to publish their own replication [(National Research Council, 2006, pp. 90-91)] of the spurious hockey stick effect M&M identified." (McKitrick, 2018)

The hockey stick is controversial, and many authorities have said it is discredited. It appeared in TAR, but in Chapter 6 of the following IPCC Fourth Assessment Report (AR4), this was said about it.

"Some of the studies conducted since the Third Assessment Report (TAR) indicate greater multi-centennial Northern Hemisphere temperature variability over the last 1 kyr than was shown in the TAR, demonstrating a sensitivity to the particular proxies used, and the specific statistical methods of processing and/or scaling them to represent past temperatures." (IPCC, 2007b, p. 436)

The list of authors and lead authors for AR4 Chapter 6 did not include Mann or Bradley; the main authors of the hockey stick. We also remember that Mann was a lead author of Chapter 2 of TAR, where the hockey stick appeared as Figure 2.20 on page 134.

Keith Briffa and the other AR4 Chapter 6 authors softened the AR4 statement quoted above as much as possible, but they said the hockey stick was wrong or at least imprecise, both in the temperature proxies chosen and the statistical methods used. A precise global temperature record, that can accurately compare the one-degree change in the past 100 years to changes over the past millennium will probably never be created. A one-degree change in 100 years is quite small. The proxies used in reconstructing past temperatures are mainly tree rings, corals, sediment records, and ice cores, plus some others as well. They all suffer, relative to thermometers, in both accuracy and temporal resolution.

The thermometers used to create the record for the past 170 years (since 1850 or so) are mostly accurate to better than ±0.5°C. Thermometer records are precisely dated and recorded daily; this is not true of proxy-temperature records. Proxy temperatures are sensitive to a particular season, usually summer, as well as precipitation levels and wind speed in addition to temperature. Tree ring proxies are also sensitive to CO_2 concentration, so modern tree ring temperatures are not comparable to ancient tree rings (Briffa, et al., 1998b). Furthermore, while thermometers are daily measurements, proxies tend to be annual to multi-annual records (Soon & Baliunas, 2003).

To make matters worse, the dates assigned to proxy datapoints are often erroneous, thus the multi-proxy averages used to construct proxy-temperature records are often not averaging values from the same time, reducing the temperature variability (or resolution) with time, making a trend or rate-of-change comparison from today to previous centuries invalid. In the words of Willie Soon and Sallie Baliunas, "The proxies used to study climatic change over the last 1,000 years are addressed individually and therefore locally because they differ in nature too greatly to be quantitatively averaged or compared." (Soon & Baliunas, 2003).

This is a very important point and often ignored. Our modern global record is reasonably accurate, especially since 2005 with the advent of ARGO[6] drifting-float ocean temperature data. The global record prior to the twentieth century is very sparse and of low quality and precision. For a particular proxy location, we can find a comparable modern temperature, but we cannot create a quality global temperature record prior to 1890–1900, the data simply doesn't exist.

The IPCC did not use the MBH99 hockey stick to show the range of possible temperatures throughout the past millennium in AR5, instead they used a variety of other reconstructions. Some are shown in Figure 2, from

---

[6] Wandering robotic floats that measure ocean temperatures from the surface to 2000 meters. Argo is named after the Greek mythological ship that Jason sailed around the Greek islands.

Chapter 5 of AR5 (IPCC, 2013). In AR6, the IPCC shows a new "hockey stick" that is suspiciously like the one in MBH99 in their summary (IPCC, 2021, pp. SPM-7).

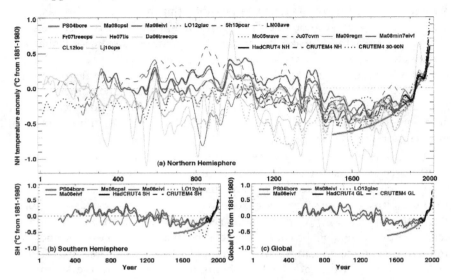

Figure 2.   Various temperature reconstructions for the Northern and Southern Hemispheres and the globe (colored lines) compared to modern instrumental temperatures (black lines). Source IPCC AR5 (IPCC, 2013, p. 409).

Several of the reconstructions show one degree or larger pre-thermometer changes in less than 100 years. Further, the range of temperature estimates in many 100-year periods are larger than one degree. One extreme example is from 1400 to 1500 AD. The graph also shows three modern high-resolution instrumental-global-temperature anomalies from the 19th century to 2000 in black. The display portrays the uncertainty in the proxy reconstructions and clearly demonstrates that one cannot definitively say the recent 0.9 degree rise in global average temperature is unusual. It could be unusual, but the data are not accurate enough to establish the fact.

The various reconstructions (the colored lines) clearly show the Medieval Warm Period (MWP) (roughly 900 to 1150 AD), which is a matter of historical record. As discussed in the IPCC caption, the reconstructions shown with red lines are land-only, those in orange are land-only extra-tropical, light blue are land and sea extra-tropical, and dark blue are land and sea all-latitudes. Prior to 1900 the number of measurements and/or proxy temperature estimates drops radically with time.

Karoly disagrees with researchers that think the MWP was warmer than today. However, the spread of values and the amplitude of the proxy temperature swings in Figure 2 show it is possible. Comprehensive historical records suggest Europe, Greenland and many other areas were warmer during the MWP, but we do not have enough data to show the rest of the world, or even the whole Northern Hemisphere, was warmer or cooler at the time.[7]

Proxy estimates of temperatures during the MWP, in Figure 2, range from -0.2° to +0.8°C, relative to today. Today means the average temperature for the past fifty years, as this is the best resolution available for the MWP. It is obviously invalid to compare a single year of instrumental temperatures to coarsely sampled proxies. We must choose a specific proxy location, then derive an appropriate modern temperature average to compare it to. Resolution cannot be increased, but it can be decreased.

Proxies are imprecise pseudo-thermometers and the global average temperature during the MWP is unknown. There are very few, unevenly spaced proxies available through the MWP; so, all we can say is that the MWP may have been warmer than today globally.

Karoly and Tamblyn on paleotemperatures

Karoly writes that the magnitude of the MWP is much greater in the North Atlantic region, in Greenland and western Europe, than in lower latitudes. The hockey stick does not wipe out the MWP but reduces its magnitude because it averages over the whole hemisphere. He is correct, peak warming, during the MWP occurred at different times in different places, likewise, the maximum Little Ice Age cooling occurred at different times in different places. Averaging temperatures or temperature proxies over an entire hemisphere or over the planet reduces the extremes, so the number of measurements, their locations, and accuracy matters.

Tamblyn also suggests that the speed of recent warming is unprecedented, but he can't be certain. The proxies have too low a temporal resolution to compute a rate of temperature change comparable to modern, daily, precise instrumental temperature records over only 100 to 150 years.

Karoly emphasizes that temperature variations in the Northern and Southern Hemispheres are often out of phase with one another in decadal and multi-decadal time frames (Karoly, 2021a, p. 20). This fact is well documented (Hannon, 2020) and discussed in more detail in the next section.

The IPCC tried to use observations twice to demonstrate that humans are driving global warming; both failed. Karoly's stratospheric "fingerprint" of human influence in SAR, suffered from cherry-picking and possible non-human causes, such as sporadic El Niños and volcanism (Daly J. L., 1997). In any case, even if the fingerprint were caused by human CO_2 emissions, the observations tell us nothing about the magnitude of human influence (May, 2020c, pp. 230-238).

The hockey stick that Mann, Bradley, and Hughes introduced in TAR failed upon closer examination of the data and the statistical methods they used. After these failures, the IPCC gave up on using observations and began using

---

[7] (Luening, Vahrenholt, & Galka, 2020)

computer models to make their case that humans control the world's climate through their greenhouse gas emissions (May, 2020c, p. 244).

The Big Picture

Both Tamblyn and Karoly assert recent warming and the rise in atmospheric CO_2 are unusual. The same assertion is made by the IPCC.[8] They relate the one part per 10,000 rise in CO_2 to warming via climate models. CO_2 is a greenhouse gas and the extra CO_2 in the atmosphere is largely from burning fossil fuels. Happer agrees with the latter point, but not with the first. That is, he does not think recent warming is unusual, or necessarily related to the rise in CO_2. The difference in their opinions revolves around climate sensitivity (CS) and the likely impact on humanity. Happer thinks CS to CO_2 is small and will be beneficial to humans. Karoly and Tamblyn think CS is high and the resulting warming is potentially dangerous. All three agree the key question in the debate is the uncertainty in climate sensitivity to CO_2.

To date, no observational evidence exists that CO_2 is the main cause of current warming and models are just models, not evidence. So, the crux of Karoly and Tamblyn's argument revolves around the assertion that recent warming is unusual. Here we deal with that idea.

Karoly and Tamblyn understand that CS estimates are not precise, but they believe the average and/or median of many diverse climate sensitivity estimates must be close to the truth. No one can be certain if a sudden (200 years or less, or since "pre-industrial") increase in global temperatures of 3°C would be dangerous. But they accept the opinions of others that it would be. There does seem to be a general acceptance that 2°C is safe, but 3°C is dangerous. These values are arbitrary. The exact year that the pre-industrial age ended is not precisely defined by the IPCC, but they often use 1750 as the beginning of the modern era, coincidentally or not, this is also the end of the worst part of the Little Ice Age in Europe, North America, and possibly China.

One-hundred years is roughly how long it would take the CO_2 concentration to double, from today (2020), in most projections. Although, the current rate of rise at 2.5 ppm/year according to NOAA's climate.gov (2017–2018), if simply linearly extrapolated, suggests it would take 160 years. Assuming a little acceleration is reasonable, as much of the world is still industrializing. The estimate also assumes unlimited fossil fuels and it is unclear if sufficient fossil fuels exist to double the current CO_2 atmospheric concentration. Jianliang Wang and coauthors estimate that the projected conventional and unconventional fossil fuel supply to 2100, if entirely burned, would only raise the atmospheric CO_2 level to 610 ppm in this century.[9]

---

[8] in AR5 (IPCC, 2013) and AR6 (IPCC, 2021)
[9] (Wang, Feng, Tang, Bentley, & Höök, 2017)

Background: Holocene temperatures—the past 12,000 years

The most recent time a temperature rise of 3°C in one-hundred years occurred was about 600AD as identified in data obtained from the high-resolution Greenland ice core record.[10] According to the Antarctic Vostok ice core record, a 2.4°C increase in temperatures occurred between 1500AD and 1600AD (Petit, et al., 1999), just as the Northern Hemisphere entered the coldest period of the Little Ice Age. Karoly has written that the Northern Hemisphere and Southern Hemisphere often change temperature in opposite directions (Karoly, 2021a, p. 20). This means, according to Karoly, that global averages mute significant regional climatic changes. Karoly didn't write this, but large regional climatic differences suggests that CO_2, a quickly dispersing and well mixed gas, is not controlling the climate.

"Global" average temperatures are a poor climate metric. Averaging temperature anomalies globally hides significant regional, natural climate forcings and assumes what is left is a global change due to CO_2. Framing the IPCC's key metric in this way forces the answer they want. Historically, and logically, climate refers to the average weather in a specific area (Koutsoyiannis, 2021). The climate humans experience is *local*, not global.

Proxies derived from ice cores typically provide better resolution and dating than most others. Over the last 1,000 years the Vostok, Antarctica ice cores reveal a temperature every 20 to 33 years, for a resolution of about 40 to 66 years (two points are required to get a rate). The Greenland cores have samples every 8 to 20 years, for a resolution of 16 to 40 years. Both should be able to resolve a 100-year change, although the accuracy of the rate is suspect for such a short period. Further, we do not know how representative the changes are in their respective hemispheres, nor do we have the precise date for each temperature. What was the year of the maximum or minimum temperature? What were the maximum and minimum temperatures in any given 100-year period? We cannot say.

In 2013, Shaun Marcott[11] collected most of the available temperature proxies that covered the Holocene Epoch—the period from 12,000 years ago to the present. They included lake-sediment, ice-core, and ocean-sediment proxies. Resolution of ice-core proxies varied from 10 years to over 500 years, averaging 136 years. Excluding the Dome F ice core from Antarctica, with a resolution of 500 years, the remaining ice cores averaged 40 years. Resolution of the lake-sediment proxies varied from 70 to 190 years, averaging around 110 years. Finally, resolution of the marine-sediment proxies varied from 40 to 530 years, averaging about 160 years.

---

[10] (Kobashi, et al., 2011) and (Kobashi, et al., 2013)
[11] (Marcott S. A., Shakun, Clark, & Mix, 2013)

Tree ring proxies, which are annual and well dated, were not included as they are unreliable due to confounding influences on their growth (National Research Council, 2006).

The Wegman Report emphasizes that the average width and density of a tree ring is a function of many variables including the tree species, tree age, stored carbohydrates in the tree, nutrients in the soil, and climatic factors including sunlight, precipitation, temperature, wind speed, humidity, and even carbon-dioxide availability in the atmosphere. The influence of carbon-dioxide concentration limits the usefulness of tree rings in modern times and makes it difficult to use modern data to predict past tree ring growth. The Wegman Report states: "tree ring proxy data alone is not sufficient to determine past climate variables." The Wegman Report also notes:

> "Tree ring density is useful in paleoclimatic temperature reconstructions because in mature trees, tree rings vary approximately linearly with age. The width and density of tree rings are dependent on many confounding factors, making it difficult to isolate the climatic temperature signal." (Wegman, Scott, & Said, 2010, p. 2)

Thus, we are limited in determining how modern temperatures compare to the past on century time scales. On land, the modern instrumental record might be reliable back to about 1900 in the Northern Hemisphere, but prior to 1950 there are very few weather stations in the Southern Hemisphere. Thermometer datasets exist for the 19th century, but coverage is sparse and of low quality, so we only have decent land data for about 120 to 125 years, and only in the Northern Hemisphere. For the oceans, we only have good sea-surface-temperature (SST) data since about 2002 to 2005. The only SST data prior to 1990 are from ships. Ship data is questionable.[12] There are very few places where we can detect pre-twentieth-century temperatures with any confidence, and most are in Greenland, Antarctica, and other high latitude locations.

Background: Rosenthal's reconstruction of Holocene North Pacific SSTs

We've already noted recent extreme warming in ice cores from Greenland and Antarctica. Very rapid warming has also been identified in seafloor-sediment cores from the Makassar Strait in Indonesia. Yair Rosenthal and colleagues, reporting in *Science*,[13] carefully examined fossil foraminifera from the cores and built a proxy temperature record for the water flowing through the strait.

---

[12] (Kennedy J. J., Rayner, Smith, Parker, & Saunby, 2011)
[13] (Rosenthal, Linsley, & Oppo, 2013)

The fossil records they examined have a 20- to 50-year resolution. The foraminifera fossils (*Hyalinea balthica*) were benthic creatures (bottom dwelling) that lived about 500 meters below the surface. Rosenthal calls this depth the IWT, which stands for intermediate water temperatures. His stated temperature error is ±0.35°C and he corrected his data for past changes in sea level. The depths that Rosenthal studied are like those from which the water temperatures are plotted in Figure 3. Rosenthal summarizes his results as follows:

> "We show that water masses linked to North Pacific and Antarctic intermediate waters were warmer by 2.1 ± 0.4°C and 1.5 ± 0.4°C, respectively, during the middle Holocene Thermal Maximum [same as the Holocene Climatic Optimum, roughly 8,000BC to 4,000BC] than over the past century. Both water masses were ~0.9°C warmer during the Medieval Warm Period than during the Little Ice Age and ~0.65° warmer than in recent decades." (Rosenthal, Linsley, & Oppo, 2013)

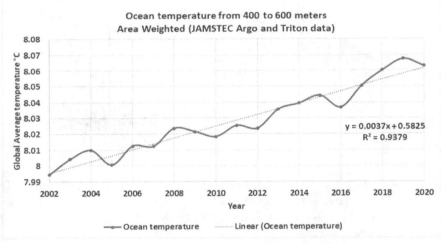

Figure 3. Modern global ocean temperatures from 400 to 600 meters. Data source: JAMSTEC.

Figure 3 is built by averaging the values populated in a global map grid created by JAMSTEC.[14] The water between 400 and 600-meters is about 8°C and is warming slowly at a rate of about 0.4°C/century. The increasing temperature is nearly linear over the period shown, with no sign of acceleration.

---

[14] The Japanese Agency for Marine-Earth Science and Technology (Hosoda S. , Ohira, Sato, & Suga, 2010)

Rosenthal's core samples were collected from sediments in the Makassar Strait, which is situated between Borneo, Java, and Sulawesi, Indonesia. The map in Figure 4 shows the 500-decibar (487-meter) water temperatures for the world ocean, using the JAMSTEC grid for 2020. The arrow on the map points to the location of the ocean-sediment cores.

The boxed area in Figure 4 is the source of the water in the Makassar Strait. In 2020, the average 500-meter water temperature in the box was 8.7°C. The overall ocean average at 500 meters was 8°C, suggesting the site was well located as Rosenthal reports in his paper. The Makassar Strait is a portion of the larger Indonesian Throughflow (ITF), a group of ocean currents that carry water from the Northern Pacific to the Indian Ocean and Southern Oceans, and sometimes vice-versa. Due to weather changes throughout the year and over decades the flow sometimes reverses in portions of the throughflow.

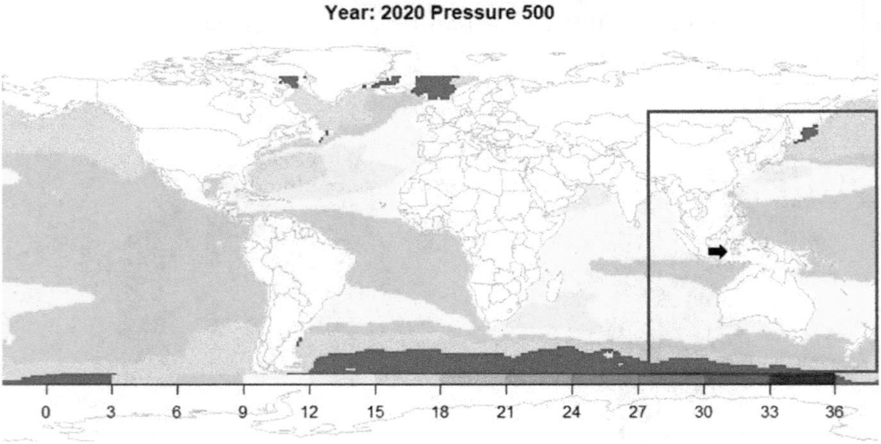

Year: 2020 Pressure 500

Figure 4. *A map of ocean temperatures at 500 decibars pressure (~487 meters). The boxed area of the Indian, Southern and Pacific oceans is the source of the waters flowing through the Makassar Strait. The Rosenthal core sites are marked with a black arrow.*

The throughflow is a major component of the global surface climate system and temperatures in these waters reflect the surface temperature of a large portion of the Northern Pacific Ocean, as well as smaller portions of the Indian and Southern Oceans. Rosenthal found that the temperature[15] at 500

---

[15] Rosenthal's original data is given as an anomaly from 1930. Here we added 7.6°C to the record, which is the University of Hamburg average temperature at 500 meters in the Makassar Strait from 2004 to 2013, to convert it into an approximate temperature. No correction from 1930 to 2009 (roughly halfway between 2004 and 2013) was attempted because the last two points in Rosenthal's reconstruction, for 1950 and 1970, suggest cooling of 0.25°C. Global temperatures did decline from 1950

meters, in the Makassar Strait, reached a maximum of ~10.7°C between 6000BC and 4000BC, the Holocene Climatic Optimum. His 500-meter temperature reconstruction is shown in Figure 5.

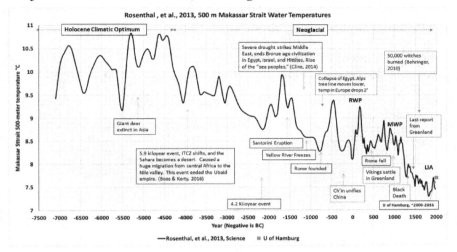

Figure 5. Rosenthal and colleagues reconstructed Makassar Strait 500-meter water temperatures. Data source: (Rosenthal, Linsley, & Oppo, 2013).

Figure 6 shows the full Makassar Strait temperature profile from Viktor Gouretski at the University of Hamburg (Gouretski, 2019). Gouretski's temperature at 500 meters is marked in red in both Figures 5 and 6. Figure 7 is a more detailed map of the Makassar Strait study area.

During the Medieval Warm Period, the 500-meter temperature reached 8.6° in 890AD; about 1°C higher than today. During the Holocene Climatic Optimum, it peaked at 10.7°C in 5,300BC; 3°C higher than today. The Roman Warm Period (RWP) was also much warmer than today in the Makassar Strait.

The Little Ice Age (LIA) was colder in the Strait than today's temperature by 0.4°C. Here the coldest part of the LIA is delayed relative to European historical records, occurring about 1810, rather than 1650–1750. We have marked some historical events on the reconstruction in Figure 5 for context. The plot emphasizes, that in the Strait, the *unusual climatic event is the LIA* or the IPCC pre-industrial period, not the modern warm period, which is still cool relative to the rest of the Holocene.

---

to 1970, but then began to rise. The UAH satellite record suggests global temperatures from 1979 to 2009 rose about 0.4°C, this is slightly more than the cooling in the Rosenthal reconstruction from 1930 to 1970, but close enough that no adjustment was required.

The coldest period in Europe was from 1650 to 1750, yet anthropogenic radiative forcing began in 1750 according to AR6.[16] The IPCC measures anthropogenic global warming from the pre-industrial period, which they define as 1850 to 1900 (AR6, SPM-5). This implies that 1850 to 1900 was some sort of ideal temperature, but as Figure 5 shows it was the coldest period in the last 9,000 years in the Northern Pacific. Historically the Little Ice Age was the coldest and most miserable time in most of the known world according to historian Wolfgang Behringer,[17] it is not a time to return to.

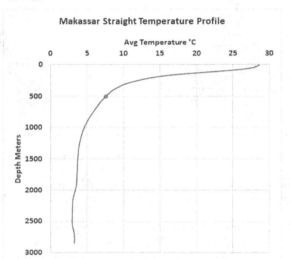

Figure 6. Makassar Strait temperature profile. The plot reflects average temperature measurements made from 2004 to 2013. The 500-meter temperature, 7.6°C, is marked in red. Data source: Viktor Gouretski University of Hamburg.

Another historian of the Little Ice Age, Geoffrey Parker,[18] tells us that one-third of the human population died because of the severe cold in the mid-seventeenth century. The seventeenth century was a period of almost continuous war from Europe to the Ottoman Empire to Russia and China.

---

[16] (IPCC, 2021, p. TS-11). All references to the latest IPCC report, AR6, in this book are from the August 9, 2021, preliminary publication. It is likely that the page numbers will change in the final version of the volume, but unlikely the content will change significantly. The IPCC has every page stamped with "Do Not Cite, Quote …," but the volume was publicly distributed and has already been quoted and cited in the *Wall Street Journal* (Aug. 9, 2021) and other publications, so we will cite it here.

[17] (Behringer, 2010)
[18] (Parker, 2012)

This was due to the cold and drought of the Little Ice Age. For the IPCC to imply the Little Ice Age had an ideal climate makes no sense.

Background: Holocene temperature varies by latitude

Figure 8 spans the entire Holocene Epoch (9700BC to the present) showing the average temperature anomaly along several latitude slices around Earth. There are large hemispheric differences in temperature and trends. The anomaly for the Northern Hemisphere, which contains most of the land area, is distinctly different from the other regions shown in the graph.

The Holocene Climatic Optimum and the Neoglacial are marked in both Figure 5 and Figure 8. The two periods are separated by the "Mid-Holocene Transition." The Neoglacial is a roughly 5,000-year period of cooling that reached its coldest point in the LIA 200 to 500 years ago, depending upon location.

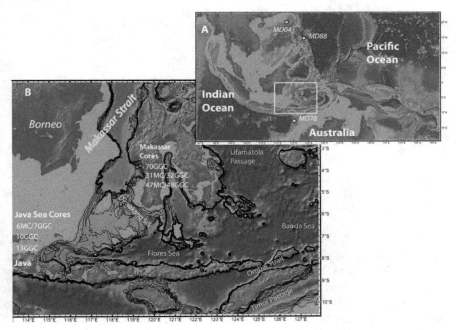

Figure 7. The location of the Makassar Strait. Source: (Rosenthal, Linsley, & Oppo, 2013)

Comparing Figures 5 and 8 we see that the Makassar Strait 500-meter record is like the Northern Hemisphere (30°N to 60°N) record in Figure 8. This is not surprising, since most of the water flowing through the Strait is from the North Pacific. In Figure 8 we see that most of the world has a much more consistent temperature from 6,000BC until the LIA. The LIA and the Neoglacial decline are clear anomalies in all regions except the Antarctic.

The Neoglacial and LIA can barely be seen in the Southern Hemisphere (30°S to 60°S) reconstruction but are quite apparent in the tropics (30°N to 30°S). The Neoglacial is subtle in the Arctic reconstruction, but the LIA is very apparent and dramatic. All this emphasizes the local nature of climate change. Constructing global surface temperature averages hides important regional climatic changes. Natural climate change tends to be by latitude, as explained by Javier Vinós in a blog post on judithcurry.com (Vinós, 2017).

CO_2 is a fairly well mixed gas and if it has a major effect on climate and surface temperature, we would expect the changes to be global. Yet, that is not what we see. Past changes are regional, not global. Much more on the reconstructions in Figure 8 can be seen in my first book.[19]

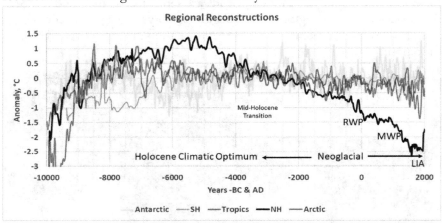

Figure 8. *Average temperature reconstructions for several latitude slices around the Earth. The Northern Hemisphere (NH) is from 30N to 60N, and the Southern Hemisphere (SH) is from 30S to 60S. After: (May, 2018)*

During the Mid-Holocene Transition, approximately 3,500 to 4,000BC, the Earth's orbital characteristics brought about a shift of the Intertropical Convergence Zone (ITCZ). The world's climate changed radically in response. It was at this time that the Sahara became a desert. The same orbital characteristics causing the shift; previously caused the end of the last glacial period. Details can be found in two blog posts by Javier Vinós (Vinós, 2017) and (Vinós, 2018), and will be discussed further in the next chapter.

The reconstructions shown in Figure 8 use a subset of Shaun Marcott's proxies.[20] We selected those that had better resolutions. Marcott's reconstructions did not preserve any temperature variability at periods shorter than 300 years, and we tried to improve on that. As a result, the reconstructions shown in Figure 8 may preserve some of the variability in periods of 200 to 250 years or more. No variability for periods less than 120

---

[19] *Climate Catastrophe! Science or Science Fiction?* Chapter 4. (May, 2018)
[20] (Marcott, Shakun, Clark, & Mix, 2013)

years is preserved. Thus, changes like the modern warm period, where temperatures increase about one degree since 1900 would not be seen.

Geological evidence uncovered by Jeffery Severinghaus and colleagues suggest that at approximately 9600 BC a rapid warming took place over the entire Northern Hemisphere.[21] At that time a step change took place and the Northern Hemisphere warmed 5 to 10°C in just several decades; or one degree per decade at minimum. Thus, the warming we have seen since 1900 was packed into ten years or less.

Contrast this with AR5, which says:

> "There is *medium confidence* that the last 30 years were likely the warmest 30-year period of the last 1400 years." (IPCC, 2013, p. 411)

Incredibly misleading. Fortunately, for the IPCC, "*medium confidence*" means roughly 50/50, or as likely as not (IPCC Core writing team, 2010). It is *likely as not* that there were multi-decade periods between 750AD and 1100AD that warmed faster than the past century and to higher temperatures.[22] S. P. Huang and colleagues use low-resolution borehole-temperature data from around the world to show that the Holocene Climatic Optimum was one to two degrees warmer than today and the MWP was about the same.[23]

From a scientific standpoint, the IPCC statement quoted above is meaningless. There are no data that span the past 1,400 years that have a 30-year resolution, either locally or globally. They have no basis for the statement. Temperature proxies do not allow decadal or 30-year temperature changes to be resolved, even as recently as 1000AD.

The Holocene climate is characterized by rapid warming as the world came out of the last glacial maximum, which occurred around 17,000BC. At that time, AR5 concluded that the globe was four to eight degrees Celsius colder than today, although the temperature difference for the Northern Hemisphere was much larger. They conclude that the maximum global warming rate during the post glacial period was one to 1.5°C per 1,000 years. Perhaps so, but the warming in the Northern Hemisphere and Arctic was much faster. While the Northern Hemisphere was warming, the Antarctic, Southern Hemisphere, and the tropics were cooling from 8000BC to 5000BC, so global averages can be deceiving.

AR5 mentions that the warming and cooling in the Northern and Southern Hemispheres are not synchronous. They also acknowledge that the differences are "fully consistent" with changes in insolation due to Earth's orbital changes (IPCC, 2013, p. 400). However, they don't discuss the extreme temperature

---

[21] (Severinghaus, Sowers, Brook, Alley, & Bender, 1998)
[22] (Christiansen & Ljungqvist, 2012) (Rosenthal, Linsley, & Oppo, 2013)
[23] (Huang, Pollack, & Shen, 2008)

swings in the Northern Hemisphere shown in Figure 8. The large differences between the Northern Hemisphere and the other latitude bands are not consistent with forcing by CO_2, a well-mixed atmospheric gas, but they are consistent, as stated in AR5, with orbital forcing. Figure 9 shows the recent lower tropospheric warming for the Northern Hemisphere, the Southern Hemisphere, the tropics, and the globe.

University of Alabama, Huntsville (UAH) satellite global-temperature records, maintained by Roy Spencer and John Christy, show that the Northern Hemisphere has warmed 15% faster than the tropics and 30% faster than the Southern Hemisphere since 1979. All are lower troposphere warming rates and are relative to the respective average from 1990 to 2020 so, their respective trends cross zero in 2005. The differences in warming rate are significant.

Global average temperatures, in general, have little meaning without a globally acting forcing agent. CO_2 disperses rapidly, so for it to be an important factor in global warming, we would expect warming to occur uniformly over the entire Earth—especially if the Sun and other factors remain static as Karoly and the IPCC claim. But this is not what happened recently or in the distant past.

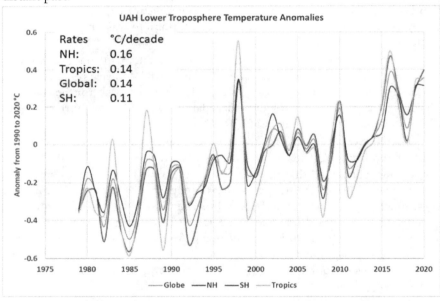

Figure 9. UAH V. 6 lower tropospheric warming rates for the Northern Hemisphere (NH), Southern Hemisphere (SH), tropics, and for the globe.

If we use surface-temperature records, the difference is even more stark. Figure 10 shows the HadCRUT5[24] temperature records for the Northern and

---

[24] HadCRUT5 is a combination dataset of HadSST version 4 sea-surface temperatures (from the Met Office Hadley Center), and CRUTEM version 5 land-

Southern Hemispheres from 1979 to the present. The records shown are a composite of land and sea measurements and cover most of the Earth.

Figure 10 suggests the Northern Hemisphere (NH) surface is warming almost three times faster than the Southern Hemisphere (SH). This is consistent with orbital forcing, which varies with latitude due to Earth's eccentric orbit and its tilt relative to the orbital plane. As with the lower troposphere warming plotted in Figure 9, this is not consistent with CO_2 forcing—which should be global—because CO_2 is a well-mixed gas. In AR5, the IPCC explains the termination of the last glacial period as follows:

> "SH temperature lead over the NH is fully consistent with the NH orbital forcing of deglacial ice volume changes (*high confidence*) and the importance of the climate-carbon cycle feedbacks in glacial-interglacial transitions." (IPCC, 2013, p. 400).

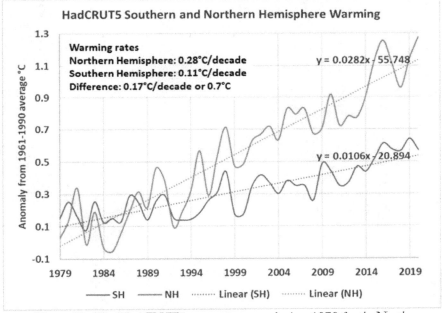

Figure 10. The HadCRUT5 temperature records since 1979 for the Northern and Southern Hemispheres and the least squares trend lines. Data source: Hadley Climatic Research Unit.

Thus, they are saying that orbital forcing terminated the glacial period, leading to the Holocene interglacial, but CO_2 released from the oceans and melting ice provided a helpful positive "feedback." This may be true, but the

---

surface temperatures (from a collaborative effort of the Climatic Research Unit at the University of East Anglia, the Met Office Hadley Centre and the National Centre for Atmospheric Science).

amount of the CO_2 feedback is unknown, as Happer makes clear. By using global as opposed to regional average temperatures, the IPCC and the consensus hide this confounding warming pattern. They frame the argument as a discussion of global averages, implicitly assuming a uniform global force, like CO_2.

The large difference in the Northern Hemisphere warming rates (0.16°C/decade versus 0.28°C/decade) between Figures 9 and 10 is hard to explain, since the warming rates for the Southern Hemisphere are the same to two decimal places. This suggests an error in one of the records, however Figure 10 is a graph of surface temperatures, and Figure 9 is for the Lower Troposphere, so while the difference in NH rates is suspicious, it is possible. Next, we investigate the geological past. How did climate change before humans arrived on Earth?

Chapter 3: What does the Geologic Past tell us?

None of the debaters discusses or presents a long-term geological reconstruction of climate, which makes understanding their discussions of the geological past hard to follow. Complex life has existed on Earth for the past 540 million years (the Phanerozoic) and reasonable, but low temporal resolution climate reconstructions of this period are available. Figure 11 is a recent one by the Smithsonian Institution that helps put our current climate into long-term geological context.

The current global average temperature is about 14-15°C (59°F), which is unusually cool. The average surface temperature over the entire Phanerozoic is about 18-20°C (a comfortable 64-68°F)—three to five degrees higher than today (Scotese, 2015). Figure 11 was created by Scott Wing and Brian Huber, with some help from Christopher Scotese. Scotese published a similar, but less extreme reconstruction in 2015 (Scotese, 2015). Their work shows that global average temperature has varied from around 11°C (52°F) to 32°C (90°F); a range of 21°C (38°F). This sort of variability is beyond anything predicted by the consensus climate models.

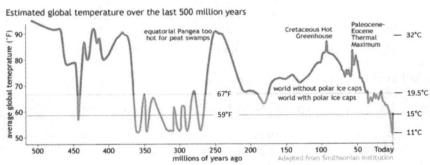

Estimated global temperature over the last 500 million years

Figure 11. Estimated global average-surface temperatures for the past 550 million years. The graph is from the Smithsonian Institution. The text discusses the Permo-Triassic Thermal Maximum at about 250 to 240 Ma, the Cretaceous Hothouse at 90 Ma, and the Paleocene-Eocene Thermal Maximum at 56 Ma. The authors are Scott Wing and Brian Huber, with help from Christopher Scotese. (Scott & Lindsey, 2020)

Figure 11 has a very low temporal resolution, perhaps five million years between the plotted values, on average (Scotese, Song, Mills, & Meer, 2021). Scotese's 2015 reconstruction has a slightly narrower range of temperatures, with peak values of 12°C and 28°C (Scotese, 2015). The resolution for the past fifty million years is somewhat better, but it is still very low-resolution.

Wing, Huber, and Scotese map the past extent of the five major Köppen climate belts using several geological climate indicators, such as coal beds (suggesting rain forest climate) and fossil glacial deposits (tillites and dropstones) that suggest glaciers and then take note of their location relative to Earth's poles at the time the rocks were deposited. They take continental drift into account and use the resulting maps to derive an equator-to-pole temperature gradient, roughly every five million years. Köppen climate belts are described very well in Scotese's 2021 review paper (Scotese, Song, Mills, & Meer, 2021).

Temperatures near the equator do not vary much over geological time, relative to changes at the poles (Scotese, 2015, p. 53). Climate change and global average surface temperature are mostly driven by what is happening in the polar regions, so by determining the gradient from the equator to the poles, global average surface temperature can be estimated. Low equator-to-pole gradients imply a warm Earth and high equator-to-pole gradients, such as we have today, indicate a cool Earth.

This method is not precise, but it gives us a good idea. The relatively small equatorial temperature changes over geological time are caused by changes in the Sun, Earth's orbit, and the location of the continental land masses. Scotese and his colleagues estimate tropical temperatures over the Phanerozoic using oxygen isotope measurements from fossils, the details are explained in their 2021 review paper.[25]

This introduction and Figure 11 are intended to help the reader understand the debater's arguments that follow. Happer points out that only one time—300 million years ago—has the atmospheric CO_2 concentration been as low as in the recent geological past (Happer, 2021a, p. 34). He suggests that we are currently in a CO_2 famine. Over most of geological history, CO_2 levels were much higher than today, but temperatures were sometimes lower.

Temperature varies with the logarithm of CO_2

Happer explains that estimated global warming over the past 170 years has been erratic; sometimes warming, sometimes cooling. He also notes in his Major Statement, that CO_2 forces warming according to the logarithm (base 2) of its concentration. This is due to the unique dependence of the CO_2 cross-section on radiation frequency (Happer, 2021b, p. 26). In Figure 12 we plot the logarithm (base 2) of the NASA CO_2 concentration since 1900 versus the HadCRUT5 global temperature record.

In Figure 12 we see Happer's point clearly. The logarithm of CO_2 and HadCRUT5 temperatures from 1978 to 2020 correlate well at the scales shown, but quite poorly from 1944 to 1976 when the world was cooling at a rate of 0.03°C per decade. The correlation is also poor from 1900 to 1944, when the world warmed faster than can be explained by the change in CO_2 concentration. Happer's point is that if CO_2 is the primary cause of current warming, why do temperature and the logarithm of CO_2 correlate so poorly?

---

[25] (Scotese, Song, Mills, & Meer, 2021)

The consensus, Karoly, Tamblyn, and the IPCC want us to believe that CO_2 causes all or most of the current global warming, but the poor correlation belies this assertion. As Happer writes, "The erratic nature of the warming over the past century suggests that half or more of the warming is not due to more CO_2 but has been caused by other natural phenomena" (Happer, 2021a, pp. 38-39).

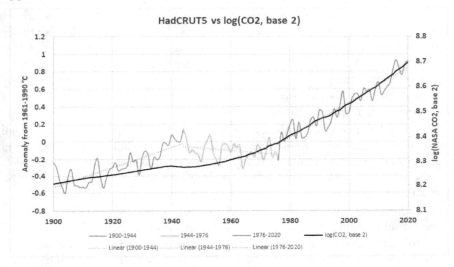

Figure 12. The HadCRUT version 5 global temperature anomaly plotted with the logarithm, base 2 of the NASA CO_2 concentration to 2011 and Mauna Loa after 2011.

The Next Glaciation

Happer points out that whatever it was that caused the recurrent glacial periods and interglacial periods of the past million years, it was not human-generated CO_2. The Earth's orbital variations affect the climate in a way described by Milutin Milankovitch in 1920 (Graham, 2000). His methodology can be used to compute the summer insolation at 65 degrees north latitude, which swings over 100 W/m² (Watts per square meter) each glacial period. Enough to initiate and stop major glacials. Glacial periods, or "glacials" are often colloquially called "ice ages."

Happer writes that these variations make much more sense than CO_2 as a cause. After all, the Antarctic ice-core records show that temperatures lead CO_2 increases by 600 to 800 years and, although the IPCC has been trying to reduce that lead (IPCC, 2013, pp. 400-401), they have not eliminated it. The lead is opposite of what would be expected if CO_2 concentration drove temperature change, and suggests it is impossible for CO_2 to cause the change (Happer, 2021a, pp. 39-40). Karoly agrees that orbital changes initiate and end glacial periods and that change in CO_2 concentration follows change in temperature (Karoly, 2021a, p. 24).

Background: how Earth's orbital variations affect climate

Here we provide a necessary bit of background on how Earth's orbit affects climate. Orbital influences, such as orbital obliquity and precession do not affect the entire globe evenly, and historically these are the largest forces acting on Earth's climate. However, the IPCC believes, if CO_2 levels remain above 300 ppm, another glacial maximum will not occur for at least 50,000 years (IPCC, 2013, p. 435). Javier Vinós disagrees, his analysis (Vinós, 2016) of past interglacial periods, including the current one, suggests they are solely dependent upon Earth's orbital characteristics, and that increasing CO_2 cannot delay or significantly modify the glacial cycle.

While Happer and Karoly agree orbital cycles caused, or mostly caused, all the glacial/interglacial transitions of the past one million years, Tamblyn is silent on the issue, but he does point out that Earth was warmer and sea levels were higher during recent interglacial periods than at present (Tamblyn, 2021a, p. 23).

Orbital eccentricity, obliquity, and precession work together in a complex way to initiate and end glacial periods. Orbital eccentricity is the variation of Earth's orbit from a perfect circle as shown in Figure 13. Periodic changes in the eccentricity of Earth's orbit are caused by the gravitational pull of the other planets. This results in changes every 413,000, 95,000, and 125,000 years. These changes vary the time at which the maximum and minimum energy strikes Earth. The result is that seasons are lengthened in one hemisphere (northern or southern) and shortened in the other. Currently the Northern Hemisphere winter is 4.6 days shorter than the Southern Hemisphere (SH). Eccentricity is a small effect. It is only through changes in obliquity and precession that eccentricity matters.

Obliquity is the tilt of Earth's axis (lower left of Figure 13) relative to the plane of the ecliptic. It affects the amount of insolation—the amount of sunlight striking Earth—at the poles during the summer versus the winter. The higher the obliquity, the higher the summer insolation and the lower the winter insolation in the polar regions and high latitudes. Higher obliquity reduces insolation in the tropics. Currently obliquity is 23.44° and decreasing. It varies from 22.1° to 24.3° throughout its 41,000-year cycle. Falling obliquity produces successively cooler temperatures in the polar regions.

Orbital obliquity affects both poles equally. Its effect on the intensity of insolation is small, but great enough to influence climate. A half-cycle from greatest to least obliquity is completed every 20,500 years and a full cycle is 41,000 years. Figure 14 shows obliquity was greatest in 7,500 BC, coincident with maximum Northern-Hemisphere (NH) precession. This was the time of greatest warming in the Northern Hemisphere.[26]

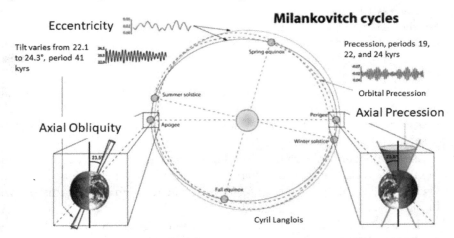

Figure 13. *The three critical elements of the Milankovitch orbital cycles, obliquity, precession, and eccentricity. After Javier Vinos and Cyril Langlois.*

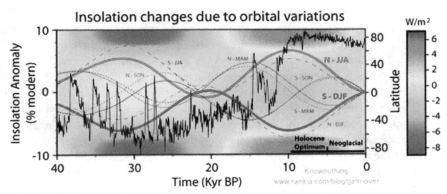

Figure 14. *Orbital insolation changes for the past 40,000 years relative to today. The Holocene Optimum and the Neoglacial are marked. The background color reflects obliquity insolation. The heavy red line shows NH-summer precession-related insolation and the heavy blue line shows SH-summer insolation. Source: Javier Vinós, used with permission (Vinós, 2017).*

Precession describes the circular wobble of the axis around a central point (lower right, Figure 13). It determines which hemisphere is facing the Sun at perihelion (closest to the Sun) and at aphelion (farthest from the Sun). The complex interaction of precession with eccentricity causes precessional cycles to occur at 14,000, 19,000, and 22,000 years. Precession doesn't change the total insolation at any latitude, but it changes the distribution by season.

---

[26] (Figure 8, Chapter 2) (Vinós, 2017)

Currently, the NH summer takes place at aphelion; thus, the northern latitudes are at the minimum (cooler summers, warmer winters) in the precession cycle.

Precession causes opposite warming effects between the Northern and Southern Hemispheres. While NH summers are warmer, SH summers are cooler, and vice-versa. At the beginning of the Holocene, Northern Hemisphere seasonality was maximal, that is, precession caused the greatest difference between summer and winter temperatures.

I will leave the details of how these three cycles initiate and end glacial periods to Vinós, who describes the process and the myths about Milankovitch orbital cycles very well. Here we will only mention that the pace of interglacials is set by orbital obliquity.

Some obliquity minimums are skipped and do not initiate or end an interglacial. Interglacial warm periods last one-half an obliquity cycle on average but are shifted 4,000 to 6,000 years due to thermal inertia, as shown in Figure 15.

Figure 8, in Chapter 2, shows that differences in warming rates between the hemispheres are the norm, not the exception. The last glacial period ended about 9700 BC (11,800 years ago). The Arctic and the Northern Hemisphere, from 30°N to 60°N warmed much more rapidly from 10,000 BC to 9000 BC than the other latitude slices shown. The reason for this is shown in Figure 14.

In Figure 14, the black temperature-proxy curve represents δ18O- isotope-ratio[27] changes from NGRIP Greenland ice core (without scale) and reflects relative atmospheric temperature. NGRIP stands for the North Greenland Ice Core Project. The precession-related insolation curves are labeled as summer, winter, spring, and fall. N (red) or S (blue) are the Northern or Southern Hemisphere, and the three-letter sequences are months. Northern and southern summer insolation are represented with the thicker curves. Background color represents changes in annual obliquity-related insolation by latitude and time, shown in a colored scale. The Holocene Climatic Optimum, and the end of the last glacial maximum, correspond to high insolation at the poles (obliquity) and rising NH insolation due to precession. The glacial period ended when both obliquity and precession were maximal in the NH.

Present-day precession is the same as during the last glacial period, but obliquity is nearly opposite. Our Neoglacial conditions represent the first 5,000 years of a 10,000-year drop into a high glacial-insolation deficit in the polar latitudes (blue polar regions in Figure 14).[28]

---

[27] (atmospheric oxygen-18 divided by oxygen-16)

[28] Vinós used source data from Pratigya Polissar (Polissar, Abbott, Wolfe, Vuille, & Bezada, 2013) and the NGRIP team (NGRIP team, 2004) to construct Figure 14.

Background: The Glacial/Interglacial structure, will CO$_2$ stop it?

The time lag between orbital forcing and ice growth or retreat suggests the orbital threshold for glacial inception is crossed thousands of years before it occurs. During the present interglacial (the Holocene), that threshold was passed long ago (Vinós, 2018). Vinós believes the next glacial inception should occur about 1,500 to 2,500 years from now, when the long neoglacial period we are now in comes to a climax and large glaciers begin to advance southward (see Figure 15). The IPCC disagrees as they estimate a much larger climate sensitivity to CO$_2$, relative to orbital forcing. The IPCC reports the following in AR5:

> "A glacial inception is not expected to happen within the next approximate 50 kyr if either atmospheric CO$_2$ concentration remains above 300 ppm or cumulative carbon emissions exceed 1000 PgC. Only if atmospheric CO$_2$ content was below the pre-industrial level would a glaciation be possible under the present orbital configuration. ... Even for the lowest RCP 2.6 scenario, atmospheric CO$_2$ concentrations will exceed 300 ppm until the year 3000. It is therefore *virtually certain* that orbital forcing will not trigger a glacial inception before the end of the next millennium." (IPCC, 2013, p. 435).

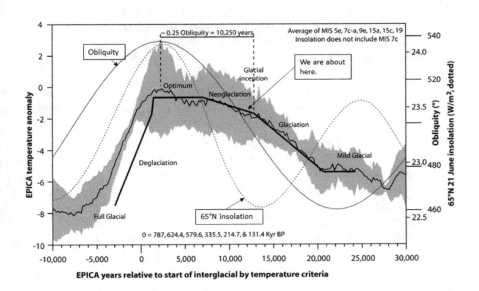

Figure 15. The elements of a normal 41,000-year glacial/interglacial cycle. The plot is an average of six of the ten most recent glacial periods. Source: (Vinós, 2018).

Vinós' orbital prediction of 1,500 to 2,500 years is compatible with the IPCC statement since the end of the next millennium (4000AD) is in the middle of Vinós' time frame. Their hypothesis that CO_2 concentrations over 300 ppm are powerful enough to hold off orbital forcing is questionable, however.

Earth's climate has been unstable and has oscillated between warm interglacial periods, such as the one we are in now, and colder glacial periods for the past 2.5 million years (Hansen B. , 2012). In his interview, Happer presents a plot of temperature and summer insolation at 65°N (Happer, 2021a, p. 39). It shows temperature variations of up to 7°C and insolation differences of up to 100 W/m². Vinós' plots similar values in Figure 15.

Moving from glacial to interglacial episodes takes about 5,000 years (the deglaciation period in Figure 15). In that time, Earth reaches a climate optimum like the Holocene Climatic Optimum marked in Figure 8 and labeled in Figure 15. Next, comes neoglacial cooling, a period of initial glacial growth, a mild glacial episode and, finally, a full glacial.

One 41,000-year obliquity cycle is shown in the figure. The right-hand scale closest to the graph is obliquity, the other right-hand scale is summer (June) insolation at 65°N. The black curve is the average of six recent interglacial-temperature-anomaly profiles. The gray area is the spread of the profiles. The heavy black line, drawn by Vinós, is broken into five straight sections with differing slopes. Each section is labeled and represents a different portion of the profile. Notice, we are currently near the end of the Neoglaciation period. The Holocene Epoch is part of the family of six interglacials used to make the graph, but only the other five were used to make the average. Four of the most recent interglacial periods, did not fit this pattern and were not included in the construction. Reasons for their exclusion are detailed in Vinós' post *The Glacial Cycle* (Vinós, 2016).

The 65°N summer insolation is about 440 W/m² in the depths of the glacial period and can be as much as 540 W/m² at the peak of the interglacial optimum. This is an enormous difference of 100 W/m², and it only applies to 65°N, and is less at other latitudes, but it terminates glacial periods.

To put this into perspective, the IPCC estimates that the total anthropogenic radiative forcing, over the industrial era, is 2.3 (1.1–3.3) W/m² (IPCC, 2013, p. 661). This is a global average, but still only three percent of the 65°N forcing that terminates or initiates Earth's glacial periods. More importantly, the change is not focused on the Northern Hemisphere, where most of the land is. There is little chance that CO_2 will put off the next ice age.

As Vinós makes clear in his posts, it is especially important to study past climate changes carefully and without bias. Treating today's climate as unique and ignoring its similarity to past events, leads to incorrect conclusions.

The largest climate drivers on Earth, as measured by the 7°C-glacial/interglacial-cycle change are the orbital cycles described above. As Vinós writes, atmospheric CO_2 concentration in the depths of the last glacial period (19,000 years ago) was around 190 ppm. It rose to 280 ppm by the 19th century, as temperatures rose six to seven degrees. If the CO_2 increase caused the warming, then its effect must have been much larger than today. As explained above, the effect of CO_2 is logarithmic, with respect to temperature. Going from 190 ppm to 280, is a 47% increase. The rise from 280 ppm to today's level of 400 is a 43% increase—nearly the same. Why would a 47% increase cause six to seven degrees of warming 12,000 years ago, but a similar increase cause less than 1.5°C of warming today as shown in Figure 12?

Andrey Ganopolski and David Archer have tried to make a case that increasing CO_2 will delay the next glacial period.[29] However, as Vinós explains, this is very unlikely, the warming effect of CO_2 is too weak.

Temperatures in the geologic past

While Karoly acknowledges that orbital variations initiate and terminate glacial periods, and that CO_2 follows temperature changes—and not the other way around—he does believe that CO_2 can drive climate change (Karoly, 2021a, p. 24). He suggests that a change in CO_2 caused the Paleocene-Eocene Thermal Maximum (PETM), a time when palm trees grew on the North Slope of Alaska and in Siberia, and forests covered Antarctica.

The PETM was the peak global surface temperature of the last sixty million years, it was accompanied by an increase of CO_2 and methane in the atmosphere. However, whether the higher temperatures caused the increases, or the reverse, is not known. At the time of the warming, CO_2 levels were likely not higher than 2,500 ppm.[30] NASA estimates the level was between 1,000 and 2,000 ppm. Some, more recent estimates are even lower, as we will see in the next section.

Tamblyn asserts that CO_2 is rising faster today than any time in the past 55 million years (Tamblyn, 2021a, p. 34). He then ominously proclaims that the only times that CO_2 rose as fast as today were during the great mass extinctions of the past. The extinction event he mentions is the end-Permian event 252 million years ago (Ma). In the Permian, it is believed that CO_2 levels were between 1,600 and 2,800 ppm or four to seven times the level today (Berner & Kothavala, 2001). However, the fossil proxies used to calculate CO_2 abundance for that time only have a ten-million-year resolution. We have no idea how quickly or slowly the Permian values changed relative to today, when CO_2 is measured many times per day. Further, we have no idea if temperatures rose before or after the CO_2 changes. The more recent data, from Antarctic ice cores, suggests that temperatures rise first, then CO_2 levels, as Happer mentioned in his interview (Happer, 2021a, pp. 39-40).

Later Tamblyn brings up the ten-million-year resolution GEOCARB III[31] CO_2 model, which is plotted in Figure 16. He criticizes Happer for pointing

---

[29] (Archer & Ganopolski, 2005), (Ganopolski, Winkelmann, & Schellnhuber, 2016)
[30] (Gehler, Gingerich, & Pack, 2016)
[31] (Berner & Kothavala, 2001)

out that the Ordovician Ice Age of 440–465 Ma occurred when CO_2 levels were 16 to 18 times higher than today. Tamblyn points out that Happer's example of Ordovician CO_2 data is not good enough to use because the resolution is so low. He seems to have the opposite view when discussing the Permian, which is only 200 million years later.

He also mentions that, during the Ordovician, the Sun was not as bright as today, so 3,000 ppm of CO_2 only has the forcing of 500 ppm today. This may be true, but the level of CO_2 in the Ordovician was higher than 3,000 ppm and possibly as high as 6,000 ppm. In any case, 500 ppm is a much higher level than today. It is also higher than the 300 ppm that the IPCC says precludes a glaciation (IPCC, 2013, p. 435). We are quite certain that glaciation took place in the Sahara and mid-Africa regions in the Ordovician, so the high CO_2 levels were ineffective in stopping it, regardless of the cause.[32] Chris Scotese estimates that the Ordovician ice cap covered more than 16 million square kilometers 445 Ma, 20% larger than the Antarctic ice cap today (Scotese, 2015, p. 47). The Ordovician ice age is identified in Figure 16, and you can see it as a blue dip in the temperature record in Figure 11 at about 445 Ma.

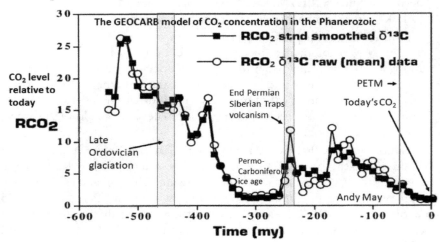

Figure 16. The GEOCARB CO_2 reconstruction for the Phanerozoic. Geological events discussed in the text have been added. After (Berner & Kothavala, 2001).

Happer also points out that atmospheric dust increased during the coldest part of glacial periods over the past 400,000 years, partially because CO_2 levels dropped so low that plants began to die from CO_2 starvation. The starvation led to more land without plant cover, increasing dust in the air.

The graph in Figure 11, and Happer's discussion in his Interview and Major Statement, make it clear that the global average temperature today is nearly the

---

[32] (Ghienne, Heron, Moreau, Denis, & Deynoux, 2009)

lowest temperature Earth's surface has seen in 600 million years. Eventually, Earth will return to its average surface temperature of 17 - 20°C, but no one knows how long it will take.

Previous climate catastrophes

We have already mentioned the Ordovician and Permian cold climate catastrophes. Both happened long before humans evolved, both were accompanied by extinction events and both saw a dip in CO_2, but the CO_2 level in the Ordovician ice age was seven times higher than in the Permian. There was an additional early cool period in the Jurassic, about 175 Ma. But the global average temperature only dropped to 16 to 17°C or so, not cold enough to be a problem.

There have been three very warm periods since the Permian, one near the end of the Permian, the second during the Cretaceous, and the third in the Early Eocene. The third is termed the "Early Eocene Climatic Optimum" (EECO) or sometimes the "Early Eocene Thermal Maximum" (EETM). The first is associated with a climate catastrophe, and all three are identified in Figure 11. The PETM, mentioned previously, occurred near the beginning of the EECO (Scotese, 2015). It is the warmest of several short, very warm periods or "hyperthermals" that occurred in the late Paleocene and Early Eocene between 56 and 53 Ma.

Background: the end-Permian warm period

Warming at the end of the Permian was also associated with a major extinction event. The extinction event itself was geologically brief and occurred between 251.94 and 251.88 Ma (Burgess, Bowring, & Shen, 2014). The extinction probably overlapped a major magmatic event known as the Siberian Traps (Black, et al., 2018). The volcanism lasted longer than the main extinction event, so it is likely that a single episode or relatively short sequence of volcanic events caused it. The end-Permian warming event began just before the extinction, but does not reach its maximum until long after, around 251.8 Ma. A maximum anoxia (oxygen deprivation) event occurred simultaneously with the main extinction and a low ocean pH event occurred afterward (Black, et al., 2018).

The Siberian volcanism began by 252.2 Ma and lasted about a million years. An extreme episode of eruptions probably occurred about 251.9 Ma that ejected huge amounts of water vapor, CO_2, and SO_2. Feifei Zhang and colleagues report that the main extinction event occurred about 251.94 Ma and was caused by an extensive, world-wide loss of oxygen (anoxia) in the oceans (Zhang, et al., 2018). The major episode was followed by possible additional anoxia events 251.7 Ma, 251.2 Ma, 250.5 Ma, and 247.2 Ma. The later episodes may or may not have caused major worldwide extinctions.

The anoxia events were probably caused by the formation of two sulfur compounds produced from reactions with SO_2 released by volcanic ejections.[33] All three sulfur compounds are toxic and could have caused both the anoxia in the oceans and atmosphere—and the extinctions. The reactions would have reduced available oxygen everywhere, but especially in the oceans. Later, the sulfuric acid would have helped to lower the ocean pH, which is what the geological record shows.

The dating of the Siberian volcanic events is not precise, but this is the best interpretation of the sequence. All the events occurred between 252.3 and 247.2 Ma. Available evidence suggests the major extinction was during a Siberian Traps magmatic event. Global temperatures probably exceeded 28°C (82°F). But Burgess and Black and their colleagues[34] suggest that the warming did not peak until more than 100,000 years after the major extinction.

The events at the end of the Permian caused the extirpation of most animal and fish species on Earth. Feifei Zhang and colleagues estimate that 90% of marine species and 75% of terrestrial species went extinct. The warming and the extinctions may have been caused by an asteroid collision, extreme volcanism in the Siberian Traps region, or both.

Background: the PETM

The rapid and extreme warming of the PETM is often cited as an example of CO_2 caused warming, so it is worthwhile to examine it in detail. The most recent warming event in Figure 11 is the PETM, about 55.6 million years ago. The PETM was almost as warm as the end-Permian event, Earth's average surface temperature rose to at least 26°C, 12°C warmer than today. Sea surface temperatures also rose at the time. One notable location was in the North Atlantic, near Denmark, where it reached 33°C.[35] Tamblyn claims the PETM caused a "small-scale mass extinction event."

The data suggests that the PETM extinctions were quite limited and accompanied by a large increase in mammalian species. Many species of benthic (bottom dwelling) foraminifera and dinoflagellates (microscopic marine animals) went extinct, but that was it. The foraminifera were mostly from the middle to deeper depths in the oceans, fewer shallower foraminifera disappeared. Other benthic animal species, such as ostracodes, living in the same environment did not show the same loss in numbers. Some speculate that the extinctions were due to greater corrosivity of deep waters, lower oxygen levels, and higher temperatures (McInerney & Wing, 2011).

---

[33] SO_2 reacts rapidly with water and oxygen to form SO_3, then SO_3 reacts with water to form sulfuric acid, (H_2SO_4).

[34] (Black, et al., 2018) and (Burgess, Bowring, & Shen, 2014)

[35] (Stokke, Jones, Tierney, Svensen, & Whiteside, 2020)

Oddly, while benthic foraminifera did not do well during the PETM, their planktonic (floating) cousins did very well, increasing in both size and diversity.

New mammals, especially primates, our distant ancestors, evolved at the time and spread widely. The PETM and EECO saw a "burst of mammalian first appearances." This period is sometimes called the "mammalian dispersal event" (McInerney & Wing, 2011). Besides mammals, the PETM and EECO saw the evolution and dispersal of numerous new and existing species of turtles and lizards. Temperatures increased six to 60°C, depending upon latitude, and there were no polar ice caps. The poles warmed the most since palm trees grew in the Arctic and Antarctica was covered in forests. Biological diversity greatly increased during the PETM, especially among terrestrial plants (McInerney & Wing, 2011). Nature blossomed then and global average temperatures were at least 12 degrees warmer than today, something to consider.

It is important to realize that nearly all mammals require a wet bulb globe ambient temperature[36] below 35°C. Above that temperature, which is roughly the mammal skin surface temperature, they cannot expel their metabolic heat. Wet bulb globe temperatures (WBGT) above 35°C[37] are fatal for most mammals after just a few hours (Sherwood & Huber, 2010). A WBGT simply means the temperature that exists when the thermometer is wrapped with a damp cloth allowing for evaporation from the bulb. A WBGT of 35 degrees C (95°F) is roughly equivalent to a heat index of 121 degrees F, which is equivalent to an air temperature of 104°F (40°C) at 42% relative humidity, on a clear day, with a wind speed of 2 mph.

Mammals not only lived during the PETM, but they also evolved and thrived. It was the time when they spread widely around the world. This means that the WBGT on land, where they lived, did not exceed 35°C, which provides an upper bound on tropical land temperatures. Mammal fossils are common during the PETM, although the most common fossil locations are in Europe, North America, and Asia north of the tropics.

The current tropics (from 23.5°S to 23.5°N) have an average ocean mixed layer temperature of about 26.7°C according to NOAA's MIMOC[38] dataset. The yearly average varies from just below 24°C to just over 28°C. During the PETM, if the maximum temperature did not exceed 35°C, the average temperature in the tropics was probably less than 34°C (Sherwood & Huber, 2010). Scotese places the average global temperature during the PETM and

---

[36] The wet bulb globe temperature is similar to the better-known heat index, but more comprehensive as it takes into account the degree of shade and wind speed (Moran, 2017).

[37] Some smaller mammals have core body temperatures of up to 38°C and can handle ambient wet bulb temperatures of 36°C (Sherwood & Huber, 2010).

[38] Monthly Isopycnal/Mixed Layer Ocean Climatology, a monthly gridded global ocean mixed layer temperature, density, and salinity dataset maintained by NOAA.

the EETM (Early Eocene Thermal Maximum) at about 25°C, roughly 10°C warmer than today.

The Cretaceous hothouse, about 90 Ma, had an average global temperature of about 28°C and Antarctic annual average temperatures were about 12°C. Mammals evolved in the Jurassic around 170 Ma; and their fossils are found throughout the Cretaceous hothouse.

The warmest time during the Cretaceous was from around 95 to 85 Ma. Like the PETM, this was also a time of rapid mammal evolution with many first appearances of fossils in the geological record.[39] Since mammals were thriving and diversifying, it is unlikely maximum temperatures above 34°C occurred during the Cretaceous hothouse.

The cause of the PETM is unknown. Recent work by Ella Stokke and her colleagues suggests that the warm period and the benthic foraminifera extinctions were closely related to, and possibly caused by, volcanism in the North Atlantic Igneous Province (NAIP). The volcanism probably caused oxygen deprivation in the Atlantic Ocean, especially in the deep Atlantic, which could have caused the extinctions.

Many theories have been proposed, including a sudden release of CO_2 and methane, but the evidence is ambiguous. It is possible that volcanism caused the release of a large amount of methane hydrates, which then caused temperatures to rise. A lot of carbon entered the atmosphere at the time, with one possible source being methane hydrates. Precipitation, at least on land, also increased (McInerney & Wing, 2011).

While total atmospheric carbon increased in the PETM, climate models have not been able to reproduce the observed large temperature increase, using estimates of CO_2 and methane at the time. The computed PETM temperatures are always too low, this makes it quite possible that the higher temperatures were not due to the greenhouse effect and had some other cause (McInerney & Wing, 2011).

NASA claims that their simulations can model the PETM temperature rise if they incorporate very high CO_2 sensitivity. Jiang Zhu and colleagues seemingly successfully simulated the PETM with existing data, but their model suggested a climate sensitivity of 6.6°C, which is not reasonable.[40] Further their model results in temperatures at or above 35°C between 30S and 30N. This is 50% of Earth's surface and at a time when mammal fossils are common in Asia, Europe, and North America. It is not impossible for mammals to have existed only north of 30N, but unlikely. One takeaway from Zhu's modeling efforts is that his models did not accurately reproduce the equator-to-pole temperature gradient in the PETM, which probably means he is missing some critical climate change component. The model overestimates tropical

---

[39] (Springer, Murphy, Eizirik, & O'Brien, 2003)
[40] (Zhu, Poulsen, & Tierney, 2019)

temperatures and underestimates polar temperatures according to available temperatures proxies.[41]

The IPCC AR6 *very likely* range of sensitivity is 2°C to 5°C/2xCO$_2$ (IPCC, 2021, pp. TS-58). In AR5, the IPCC is more explicit: "ECS is *very unlikely* greater than 6°C is an expert judgment informed by several lines of evidence." (IPCC, 2013, p. 1111). AR5 follows with a list of the evidence why ECS is not greater than six degrees.

Tamblyn makes quite a point about how the "rate of rise of CO$_2$ levels today is unprecedented in at least the last 55 million years." He also proclaims that "CO$_2$ levels today are rising *10 times faster* than during the PETM" (Tamblyn, 2021a, p. 35). We are not sure how he computed this number, and he does not explain it.

Between 55 and 56 Ma, there are only 16 proxy estimates of CO$_2$ levels. According to a summary paper by David Beerling and Dana Royer in *Nature Geoscience*, the samples were taken at seven unique times, averaging one sample every 143,000 years. These seven dates are plotted in Figure 17. Dating error for the samples is estimated to be ±500,000 years and error in the CO$_2$

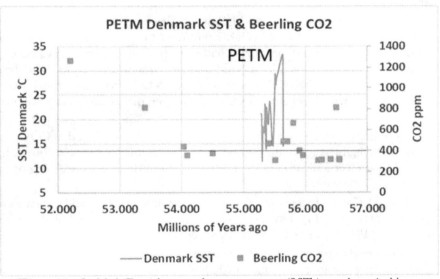

Figure 17. Stokke's Danish sea surface temperatures (SSTs) are shown in blue and Beerling's CO$_2$ estimates are shown with orange boxes. Today's global average temperature and CO$_2$ concentration are identified with the horizontal blue line.

---

[41] (Zhu, Poulsen, & Tierney, 2019) and (Scotese, Song, Mills, & Meer, 2021). This modeling problem has been named the "equable climate problem" or the "low gradient problem." (Huber & Caballero, 2011). Huber and Caballero make two things clear; models of the Eocene climate do not match the data we have and the data we have are uncertain.

estimates is ±235 ppm (Beerling & Royer, 2011). Stokke's Denmark SSTs are also plotted in Figure 17 for comparison.

AR6 claims, with *low to medium confidence*, "that the millennial rates of CO_2 concentration change in the atmosphere during the last 56 Myr [millions of years] were at least 4–5 times lower than during the last century." Then in the very next sentence, they say there is "*high confidence* that the rates of atmospheric CO_2 and CH_4 change during the last century were at least 10 and 5 times faster, respectively, than the maximum centennial growth rate averages of those gases during the last 800 kyr" (AR6, page 5–18). In the quote, "kyr" means thousand years ago. Clearly, there are no data available with the temporal resolution and accuracy required to support either statement.

Currently, we measure atmospheric CO_2 many times a day and each sample has a precise date and time. The data from 56 Ma, are quite obviously not adequate to determine a rate comparable to today.

Beerling and Royer's CO_2 data suggest that the CO_2 concentration 55.6 Ma was only 487 ppm (328–667 ppm). This is somewhat lower than other estimates and slightly higher than today, but global average temperature was at least 12 to 14°C higher.

Beerling and Royer claim the highest CO_2 level occurred 52 Ma, over four million years after the PETM. They have two CO_2 estimates from that time: 1,868 ppm (1,092–3,501 ppm) and 659 ppm (439–878 ppm). Elevated CO_2 levels exist from 54 to 32 Ma and average around 800 ppm. The peak temperature was reached earlier at 55.6 Ma, when CO_2 levels were much lower. The lack of correlation between CO_2 and temperature is easily seen in Figure 17.

During the warmest part of the PETM, we can be confident that CO_2 levels were only slightly higher than today, not high enough to be a factor in the warming. We certainly do not have any idea about the rate of warming or the rate of CO_2 increase.

The IPCC want to use the PETM as an example of what can happen today (IPCC, 2021, p. 5-14). However, the evidence suggests that life flourished during the PETM, and conditions were very different from today.

They are also trying, without success so far, to model it. How accurate are their models? We look into that in Chapter 5, next let us look at the potential influence of the Sun.

Chapter 4: The Sun

> "For good ideas and true innovation, you need human interaction, conflict, argument, debate." Margaret Heffernan, CEO and author.

Happer believes that both variations in Earth's orbit, like we discussed in Chapter 3, and variations within the Sun affect our climate more than the IPCC and the consensus will admit. In his interview he notes the striking correlation between sunspot records and well-documented historical climatic changes (Happer, 2021a, p. 41). More sunspots suggest a more active Sun, and fewer suggest a quieter Sun.

Happer also refers to recent research that shows sunspot frequency is related to the production of Carbon-14 (^{14}C) in the atmosphere.[42] This isotope is created by galactic cosmic rays. Carbon-14 concentration can be measured in tree rings and when solar activity—as measured by sunspot counts—is high, ^{14}C is low and vice versa.[43] The ^{14}C record also correlates well with historical climate changes. When ^{14}C counts are low, the climate is warm and when high, the climate is colder. Henrik Svensmark and Nir Shaviv have written extensively about this.[44] They believe that a high cosmic-ray count causes more low-level clouds, which cool the climate. They also believe that high solar activity produces a stronger solar magnetic field which protects Earth from cosmic rays, warming the Earth.

As we discussed above, Karoly and Happer agree that Earth's orbital fluctuations initiate and end glacial periods. However, while Happer believes that solar variability affected climate over the past few hundred years, Karoly does not.

The debate over recent solar activity and its effect on global warming is important (Connolly et al., 2021). To compute the human contribution to global warming, the IPCC first uses a model to predict what the temperature would be without humans; a "natural-forcings only" temperature record. Then they compute a model with both natural and human forcings; an "all-forcings" model. By subtracting the natural-forcings from the all-forcings model, they derive the human component (IPCC, 2013, pp. 879-884). The process is illustrated in Figure 27(b) and discussed in the next chapter. Assumptions made about the level of solar activity directly affect the IPCC computation of the impact of human greenhouse gas emissions.

---

[42] (Stuiver & Quay, 1980)

[43] (Svensmark, 2019, pp. 4-5)

[44] (Svensmark, Influence of Cosmic Rays on Earth's Climate, 1998) and (Svensmark, Force Majeure, The Sun's role in climate change, 2019) (Shaviv, The Spiral structure of the Milky Way, cosmic rays, and ice age epochs on Earth, 2003b)

<u>Background: The Schwabe solar cycle and historical events</u>

Nicola Scafetta and Fritz Vahrenholt point out numerous correlations between climatic events and the basic ~11-year Schwabe solar cycle (Scafetta & Vahrenholt, 2022-in press). They point to references that show the solar cycle is seen in tree rings in southern Germany and Japan, in Ionian and Bering Sea sedimentary deposits, in temperature records of Portugal and Vancouver Island, in the Indian monsoons, in the North American, European, and Arctic winter climate, in the Mexican Pacific coast cyclone frequency, in the Pacific Walker Circulation, in the oceanic heat content of the upper 700 m of the Pacific, in the thunderstorms of Brazil, and in the water vapor concentration over the Arabian Peninsula.

Longer-term solar quasi-cycles are also seen in Earth's climate. Scafetta and Vahrenholt cite sources that show the solar Gleissberg cycle (~90 years) and Suess-DeVries cycle (~210 years) are observed in the Atlantic deep water circulation, in the westerly winds of the Falkland Islands, in the climate of the Northeast Pacific, in the South American monsoon of Northeast Brazil, in the temperatures of Tibet, in the precipitation of China, in the nitrate content of the polar ice caps, in the growing season of the Northern Hemisphere, in the subtropical monsoon of the Northern Hemisphere, and in global tree ring data.

This is only a partial list of examples. In short, there is abundant evidence that solar quasi-cycles affect Earth's climate. We do not understand the way the Sun affects our climate, but it clearly does. It is not an area of study the IPCC should ignore.

Atmospheric fingerprint

Karoly believes that the change in solar radiation over the past 60 years nets to zero and has no trend beyond the normal ~11-year solar cycle. This is also the opinion of the IPCC. Karoly writes that we can confidently exclude the Sun as a contributor to recent warming, because, if it were, the stratosphere would be warming. Instead, it is cooling (Karoly, 2021a, p. 24). Recall from Chapter 2, cooling of the stratosphere when the troposphere warms is sometimes called the atmospheric fingerprint of human-caused global warming, a concept in which Karoly is an expert.

<u>Background: the atmospheric fingerprint</u>

Cooling of the stratosphere due to a higher concentration of CO_2 is logical. CO_2 is a major radiator in the stratosphere, and more of it should help cool that layer. However, the idea that this *proves* CO_2 concentration controls the climate or is responsible for more than 50% of recent warming is very controversial. There are alternate explanations for the observed cooling (Daly J. L., 1997). A short article by Neville Nichols published in *Nature*, in 1996, provides a fair summary of the atmospheric fingerprint controversy that is still accurate today. A critical portion is quoted here:

> "The study of Santer et al., and those reported in the IPCC Second Assessment, show that an anthropogenic component of global climate change—the 'anthropogenic fingerprint'—may be appearing in the observed data. It must be pointed out, however, that this signal is the complicated pattern of change resulting from the combined effects of stratospheric ozone depletion and increased concentrations of greenhouse gases and sulphate aerosols. It does not mean that the effect of any one of these factors has been detected.
>
> "Many uncertainties remain in this work and are acknowledged by Santer et al. and in the IPCC Second Assessment. There are uncertainties in estimates of the magnitude and patterns of the various natural and human-induced factors likely to affect climate. Climate models are far from perfect." (Nicholls, 1996)

As Nicholls implies, if the intensity of solar radiation (TSI) increased, ozone in the stratosphere should increase (NOAA, 2010). The ozone would absorb more ultraviolet (UV) radiation, which should warm the stratosphere, not cool it. But, whether due to human CFC emissions or due to changes in the Sun, ozone has decreased in the stratosphere, decreasing stratospheric warming. In summary, there are too many alternative explanations for the cooling in the stratosphere to draw any conclusions from it.

Satellite-temperature trends in the lower stratosphere, middle troposphere, and the lower troposphere are shown in Figure 18. Different factors drive temperatures within each layer, thus there are large excursions between the trends prior to the mid-1990s and after. Since the mid-1990s, the stratosphere and middle troposphere have had the same pattern. They are cooling while the lower troposphere is warming. Part of the pattern could be human CO_2 emissions; but can we be certain? In a word, no.

Benjamin Santer wrote a model-based paper published by *PNAS* in 2013 (Santer, et al., 2013). He acknowledged that their natural-forcings models do predict cooling in the stratosphere and warming in the lower troposphere— just as their all-forcings models do—but the match to observations was not as good. Santer used the IPCC's RCP8.5-emissions scenario to complete the study. It describes a worst-case scenario that has since been discredited by Wang and colleagues[45] and, therefore, is considered implausible.

As shown in Figure 18, the middle troposphere warms from the middle 1970s to the middle 1990s, and then cools until 2020. The lower stratosphere has the opposite trend from the 1970s to 1995 and then begins to track the

---

[45] (Wang, Feng, Tang, Bentley, & Höök, 2017)

middle troposphere. The lower troposphere tracks the middle troposphere until about 1995, then continues to increase, while the middle troposphere begins to cool. The CO_2 trend shows no change in 1995, as shown in Figure 12; clearly other factors, presumably natural, are at play. We will remember that the enormous Mt. Pinatubo volcanic eruption occurred in 1993.

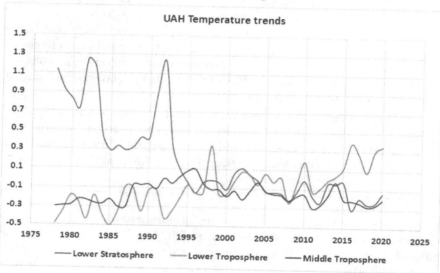

Figure 18. UAH satellite temperature trends in the global lower stratosphere and lower and middle troposphere.

Santer's natural plus anthropogenic-forcings model shows the middle and lower troposphere warming rapidly from the mid-1990s on, as shown in Figure 19 in red. He has modeled the Mt. Pinatubo eruption, and this causes a drop in the lower and middle tropospheric temperatures. His natural-forcing-only model (blue curves) shows both flattening in 2000, after the effects of the Mt. Pinatubo eruption wear off (Santer, et al., 2013, Fig. 1). Neither model fits the observations shown in Figure 18. Thus, his idea that the atmospheric temperature profile is proof of anthropogenic warming appears to be invalid.

We agree it is logical to expect additional CO_2 to warm the lower troposphere and cool the stratosphere, but logical speculation does not confirm Santer's model. Why does his model get the middle-troposphere temperature trend wrong after 1995? Compare Figure 19(B) to Figure 18. We can see the impact of Mt. Pinatubo in the observations and in the models, but the trends do not match. The reason or reasons are unknown.

Clearly, how warming and cooling are distributed in these three sections of the atmosphere is complex and poorly understood. It cannot simply be the result of human influence. If the vertical temperature profile contains an anthropogenic signal, it has not yet been successfully detected, modeled, or measured. Santer claims his results are statistically significant, but to believe this, we must accept his assumptions that the models are robust and the Sun and other natural forcings are insignificant. As we will see in the next chapter, these assumptions are probably incorrect.

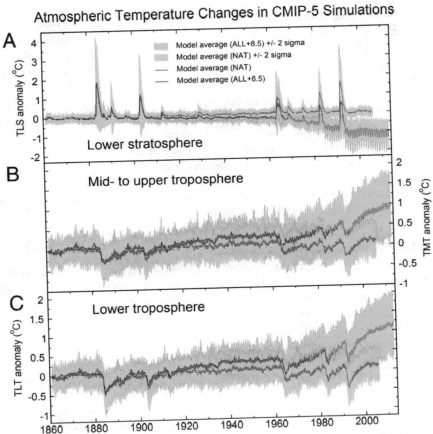

Figure 19. Santer's models of A) stratosphere B) middle troposphere and C) lower troposphere. Red is natural plus anthropogenic, and blue is natural only. The dips in the troposphere and the peaks in the stratosphere are due to modeled volcanic eruptions. Source: (Santer, et al., 2013)

TSI and the IPCC

Karoly's other argument against the Sun's influence on recent warming is that, if the warming were caused by solar changes, we would expect it to be more intense during the day and in the summer. Yet, we see the opposite. Warming is occurring more at night and in the winter. Night and winter warming are also consistent with more cloud cover and more cloud cover is expected as the world warms. Karoly's point is valid, but, like the atmospheric fingerprint, there can be natural causes that produce the same effect.

The Sun does not only vary in its radiation output, or total solar irradiance (TSI), the Sun's particle output, known as the solar wind, also varies as does the intensity of its magnetic field. Further, TSI encompasses all wavelengths of radiation and, as we will see, the wavelength—or its inverse, frequency—matters a lot in how radiation affects the climate. Higher frequency, more energetic, UV radiation from the Sun varies much more than the total.[46] All these variations affect Earth's climate in ways that are poorly understood and vigorously debated.

Further, while the roughly eleven-year solar cycle exists (and is one of the few things all astrophysicists agree on), the long-term variation of the Sun is in dispute. David Karoly, the IPCC, and other scientists such as Judith Lean, believe there is no long-term trend, at least over the last 300 to 400 years. They assume it is essentially flat, or possibly declining slightly. Happer, and many astrophysicists, such as Willie Soon and Nicola Scafetta, think there is a long-term, or secular, trend and that solar variation plays, or could play, a large role in global warming (Connolly et al., 2021).

Background: measuring solar output

Measuring the output of the Sun is challenging. The Sun emits so much radiation that satellite sensors begin to deteriorate as soon as they are pointed at it. Numerous designs have been tried and all have had problems maintaining their accuracy for extended periods. Splicing these records together to create one high quality record of solar output is very difficult. While AR4 was in preparation, the accepted TSI-activity record, from satellite measurements, was the Active Cavity Radiometer (ACRIM) composite. It was built by Richard Willson and his team (Willson, 1997). Scafetta and Willson reported in 2014:

> "Our analysis provides a first order validation of the ACRIM TSI composite approach and its 0.037%/decade upward trend during solar cycles 21–22 [1986–1997]. The implications of increasing TSI during the global warming of the last two decades of the 20th century are that solar forcing of climate change may be a significantly larger factor than represented in the CMIP5 general circulation climate models." (Scafetta & Willson, 2014)

Sallie Baliunas and colleagues, in 1995, published evidence in *The Astrophysical Journal* that the IPCC assumptions of the Sun's long-term variability are lower than that of similar stars (Baliunas, et al., 1995). So Scafetta and Willson's idea has support in the existing literature.

---

[46] (Krummheuer & Krivova, 2015)

Judith Lean,[47] was the lead author in charge of the relevant section of AR4.[48] She led the development of a rival TSI composite, called PMOD.[49] It shows solar activity as static or declining from 1986 to 1997. The IPCC AR5 report agrees with Lean's PMOD model and states that a declining trend in solar activity is *"very likely"* (IPCC, 2013, p. 689). Claus Fröhlich and Judith Lean concluded:

> "These results indicate that direct solar total irradiance forcing is unlikely to be the cause of global warming in the past decade, the acquisition of a much longer composite solar irradiance record is essential for reliably specifying the role of the Sun in global climate change." (Fröhlich & Lean, 1998)

Fröhlich and Lean conclude that TSI is unlikely to have caused any global warming, then say they do not have enough data to be sure. As we will see, the two groups used different techniques to reach their respective conclusions. The ACRIM group used an engineering approach and examined the raw data from satellites. The PMOD group simply compared the computed TSI results from satellites to a model of what they thought the satellite readings should be. Using the latter approach, it is no wonder that they want a longer record.

Background: ACRIM v. PMOD

Building the ACRIM and PMOD composites is complex because the satellite measurements must be scaled properly to fit together end-to-end.[50] Whether the ACRIM or the PMOD composite is used to calibrate solar proxies matters.[51] Nicola Scafetta and colleagues[52] compared the two composites and provided evidence that the ACRIM composite should be preferred.

It is an important controversy, yet it appears the decision to ignore the ACRIM composite and the more active TSI reconstructions was political. In a 2003 interview with NASA's Rebecca Lindsey, Lean explains:

---

[47] Senior Scientist for Sun-Earth System Research at the U.S. Naval Research Laboratory

[48] AR4, Chapter 2.7, p. 188, "Natural Forcings"

[49] PMOD is a solar irradiance composite from the *Physikalisch-Meteorologisches Observatorium Davos* from where it gets its name (Fröhlich & Lean, 1998).

[50] The process is discussed in (Scafetta & Willson, 2014), (Scafetta, Willson, Lee, & Wu, 2019), (Fröhlich and Lean, 1998), and (Kopp & Lean, 2011).

[51] (Fröhlich & Lean, 1998)

[52] (Scafetta, Willson, Lee, & Wu, 2019)

"The fact that some people could use [the ACRIM group's] results as an excuse to do nothing about greenhouse gas emissions is one reason, we felt we needed to look at the data ourselves. Since so much is riding on whether current climate change is natural or human-driven, it's important that people hear that many in the scientific community don't believe there is any significant long-term increase in solar output during the last 20 years." (Lindsey, 2003)

The ACRIM and PMOD TSI reconstructions are shown in Figure 20. The red straight line in each graph is at 1,360.62 W/m². The ACRIM record shows an increase to 1996–1997 and a long-term increasing "secular" trend, which continues until about 2000. It then begins to slowly decrease, coinciding with the "Pause."[53] The pause in global warming, from roughly 2000 through 2013, is discussed in more detail in Chapter 7. This is best seen by comparing the solar-cycle minima. The PMOD composite shows a pronounced and steady decreasing trend. Besides coinciding with the Pause, the ACRIM secular trend is also reminiscent of the middle-troposphere-temperature trend shown in Figure 18.

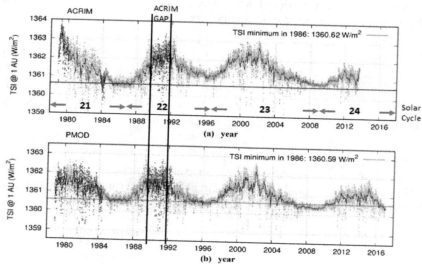

Figure 20. The ACRIM and PMOD TSI reconstructions. The trend of the ACRIM composite is up, until about 2000 and then down. The trend of the PMOD composite is flat to down. Source: Modified after (Scafetta, Willson, Lee, & Wu, 2019). Used with permission.

---

[53] (Scafetta, Willson, Lee, & Wu, 2019)

The Coupled Model Intercomparison Project (CMIP) modelers require a long-term TSI reconstruction as one of their input datasets. These reconstructions use sunspot records and many other proxies of solar activity calibrated to modern satellite measurements to extrapolate solar output as far back as 1700AD.

The most significant difference between the two is the overall TSI trend from 1986 to 1997—years of the minima preceding and following Solar Cycle 22. The reason is how they handle the so-called "ACRIM gap." The ACRIM gap came about because of the Challenger disaster of 1986. The ACRIM 2 instrument was supposed to be shuttle launched before ACRIM 1 went silent in 1989, but the disaster delayed ACRIM 2 for two years, and as a result, there was no functioning high-quality TSI-measuring satellite from mid-1989 until late 1991 (Willson, 2014). Only the Nimbus7/ERB and the ERBS/ERBE satellites were functioning, and their measurements displayed opposite trends. The Nimbus7/ERB measurements trended up 0.26 W/m² per year and the ERBS/ERB trended down 0.26 W/m² per year.[54] The difference was enough that one of the satellite instruments had to be wrong.

The PMOD group attempted to show there were problems with the Nimbus7/ERB instrument by using solar-proxy models. Then they significantly altered the instrument's TSI measurements within the ACRIM gap, including reversing the slope from positive to negative (Fröhlich & Lean, 1998). Further, they modified measurements from both ACRIM instruments, claiming they had sensor problems. The PMOD group's changes were made without consulting the original teams of scientists assigned to the satellites and without examining the raw data. Their idea was that their solar-proxy models were superior to the data and could be used to "fine-tune" the observations.[55]

The PMOD corrections to the satellite TSI data were based upon a model of solar activity built using detailed sunspot and faculae (solar bright spots) records and the assumption that sunspot characteristics and their relationship to TSI has remained fairly constant.[56] Regarding the "corrections" the PMOD team made, the leader of the Nimbus7 team, Douglas Hoyt, wrote:

> "[The NASA Nimbus7/ERB team] concluded there was no internal evidence in the [Nimbus7/ERB] records to warrant the correction that [PMOD] was proposing. Since the result was a null one, no publication was thought necessary. Thus, Fröhlich's PMOD TSI composite is not consistent with the internal data or physics of the [Nimbus7/ERB] cavity radiometer." (Scafetta & Willson, 2014, Appendix A)

---

[54] (Scafetta, Willson and Lee, et al. 2019)
[55] (Scafetta, Willson, Lee, & Wu, 2019) and (Fröhlich & Lean, 1998)
[56] (Connolly et al., 2021) (Fröhlich & Lean, 1998)

In Lean's 1995 paper in *Geophysical Research Letters*:

> "Deviations of the SMM and UARS data from the reconstructed irradiances in 1980 and 1992, respectively, may reflect instrumental effects in the ACRIM data, since space-based radiometers are most susceptible to sensitivity changes during their first year of operation." (Lean, Beer, & Bradley, 1995)

Yes, Lean is saying that her models *may reflect* that the instruments are wrong. Modifying measurements to match an unvalidated model is not accepted scientific or engineering practice. Besides the original corrections to the satellite measurements made by the PMOD group, new, and different, corrections were later suggested by Claus Fröhlich (Fröhlich, 2003). Which set should we use? Scafetta and colleague's comment:

> "a proxy model study that highlights a discrepancy between data and predictions can only suggest the need to investigate a specific case. However, the necessity of adjusting the data and how to do it must still be experimentally justified. By not doing so, the risk is to manipulate the experimental data to support a particular solar model or other bias." (Scafetta, Willson, Lee, & Wu, 2019)

The ACRIM group and Douglas Hoyt believe the upward trend in the Nimbus7/ERB data is more likely correct than the modeled downward trend created by the PMOD group. Further, the Nimbus7/ERB trend is supported by the more accurate ACRIM 1 instrument. The downward trend of the ERBE instrument is in the opposite direction of the ACRIM trend and was caused by well-documented degradation of its sensors. The ACRIM team investigated the PMOD corrections to the ACRIM 1 and ACRIM 2 data and found that they were not justified.

The ACRIM trend is up until about the year 2000 and then reverses. We don't know when the uptrend began, but it could have contributed to the warming seen in the latter part of the 20th century. The PMOD trend is flat or trending downward over the entire satellite record. If you believe the PMOD trend, then the IPCC conclusions as well as Karoly's make sense. However, there appears to be more evidence that the ACRIM trend is the correct one.

Background: Reconstructing the Solar Past

The choice of which TSI composite to use, either ACRIM or PMOD, affects solar-output reconstructions significantly (Connolly et al., 2021). Connolly, et al. 2021 show several examples of 19th and 20th century solar reconstructions and how they differ depending upon the choice of ACRIM or PMOD for calibration.

The choice of which solar proxies to use is also important. Solar proxies, such as sunspot records, carbon-14, chlorine-36, and beryllium-10 isotopes are used to extend estimates of solar variability far into the past (Haigh, 2011).

The IPCC favors, and recommends[57] their modelers use, the Wang and colleagues 2005 reconstruction shown on the upper right of Figure 21[58] and all the AR5 models used either Wang's or one similar. It had the effect of attributing nearly all warming to humans.

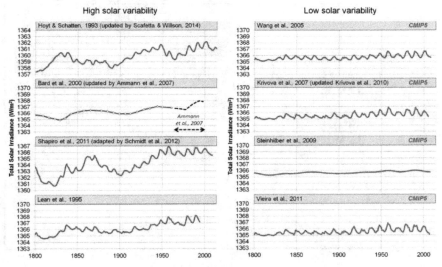

Figure 21. *Various peer-reviewed models of solar output over the past 200 years. The IPCC uses a low variability model, like those on the right, in their calculations of human influence on climate and ignores the high variability models shown on the left. See: (Soon, Connolly, & Connolly, 2015) for a full discussion of the models. Note all charts are scaled the same.*

Willie Soon, an astrophysicist with the Harvard-Smithsonian Center for Astrophysics, favors a reconstruction by Hoyt and Schatten as modified by Scafetta and Willson. It is shown on the upper left of Figure 21.[59]

The Wang, 2005 reconstruction shows less variation in solar output than most of the others and its underlying long-term trend is flat. A similar, but newer reconstruction, by Greg Kopp and Judith Lean (Kopp & Lean, 2011) shows even less. It is compared to the Hoyt and Schatten reconstruction in Figure 22.

---

[57] (Connolly R. , 2019)
[58] (Wang, Lean, & Sheeley, 2005)
[59] (Soon, Connolly, & Connolly, 2015)

The climate models used for IPCC's AR4 report (IPCC, 2007b) were from the Coupled Model Intercomparison Project, Phase 3 (CMIP3). Of five "natural-only" CMIP3 models used in the report, four used the low-variability solar reconstructions favored by Lean (Connolly R. , 2019). So, it is unsurprising that the natural-forcings model does not match the obviously natural warming from 1910 to 1944 (see Figure 12) and the natural-only models, discussed in Chapter 5, show no warming since 1951. The climate models used for AR5 also used solar reconstructions calibrated to the PMOD composite.

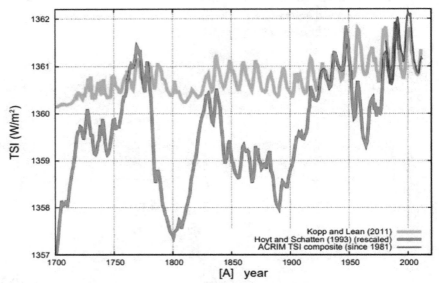

Figure 22. The Hoyt and Schatten TSI reconstruction, calibrated to the ACRIM TSI composite, is shown in red. The Kopp and Lean reconstruction, calibrated to the PMOD composite, is shown in green. The ACRIM TSI composite is in blue. Source: (Scafetta & Willson, 2014) figure 16. Used with Nicola Scafetta's kind permission.

The more active reconstructions, on the left of Figure 21, use more solar proxies than the "quieter" ones on the right. The quieter reconstructions rely heavily on sunspot numbers.[60] Thus, when there are no sunspots, no solar variation is shown. Yet, in periods of no sunspots, there are still variations in solar activity. This causes both the total and the long-term variability on the right-hand constructions to be underestimated. See Figures 23, 24, and 25 to view this comparison for solar cycle 24. Numerous researchers have found that variations in sunspots and solar faculae alone cannot account for the total solar variability (Connolly et al., 2021).

---

[60] (Soon, Connolly, & Connolly, 2015)

By ignoring the more active TSI reconstructions, the IPCC have not considered a major source of uncertainty. Both the ACRIM- and PMOD-based reconstructions should have been used, or the reason for rejecting the ACRIM composite altogether explained to everyone's satisfaction. In addition, the use of more solar proxies, especially those that show variability when there are no sunspots, should have been considered.[61] As Annie and Edward Walter Maunder wrote in 1908, "sun-spots are but one symptom of the sun's activity, and, perhaps not even the most important symptom." (Maunder & Maunder, 1908, pp. 189-190).

Two modern reconstructions, one from each camp, are compared in Figure 22, with both extended back to 1700AD. The difference between the two is significant and striking. The Kopp and Lean reconstruction varies 2 W/m² and the Hoyt and Schatten reconstruction varies 5 W/m². After correcting the difference of 3 W/m² for Earth's albedo and the fact that Earth is a sphere, the difference is still 0.5 W/m² at Earth's surface or 23% of the total IPCC AR5 estimated forcing (warming) due to humans from 1750 to 2011, about 2.3 W/m² (IPCC, 2013, p. 696). In AR6, for the period 1750 to 2019, the estimate of human forcing was increased to 2.72 W/m² (IPCC, 2021, p. 7-50).

Thus, the direct forcing, via solar radiation striking the surface of the Earth, is small, but that is only one possible way for solar variability to affect Earth's climate. As already mentioned, changes in the solar magnetic field and the solar wind also have an effect. The direct solar forcing, through TSI, can also be amplified by processes that operate on Earth. Earth does not radiate energy evenly and much of the energy it radiates to space is from the winter pole, especially the North Pole.

The north polar vortex is weaker than the southern, so thermal energy is more easily transported to it from the tropics. In the north polar winter, the specific humidity is very low, and the pole receives no solar radiation. The greenhouse effect is very weak due to the lack of water vapor and the sky has few clouds, so it functions as a powerful heat radiator to space.[62] Because the northern polar vortex is weaker, the transport of heat from the equator to the North Pole can be modulated by the Quasi-Biennial Oscillation (QBO), ENSO, and solar activity.

The QBO is a stratospheric wind that blows in an easterly direction for a period, dies down, and then is replaced by a westerly wind. When it is easterly it is called QBOe and when blowing westerly, it is called QBOw. Karin Labitzke has shown that during the QBOw phase, the North Pole winter temperatures are strongly dependent upon the sunspot number, or the strength of the Sun. This dependence is not as apparent for QBOe. Solar strength also affects ENSO, which influences the amount of heat transported

---

[61] (Scafetta & Willson, 2014) and (Soon, Connolly, & Connolly, 2015)
[62] (Marshall, 2018) See section 2, 6x10¹⁵ Watts of energy is the implied transport.

to the winter pole.[63] These and other solar amplifiers and/or modulators enhance the solar effect on Earth's climate.

Based upon model results from IPCC's AR4 and AR5 reports, "more than half" of all recent warming is attributed to humans. Both sets—the natural-forcings and all-forcings models—assume that solar variability was net zero throughout the period, which is unlikely.

What would the result be if the Hoyt and Schatten reconstruction were used? Soon, et al.[64] did the calculation for the Northern Hemisphere. They simply correlated the various solar reconstructions in Figure 21 to global average temperature reconstructions to examine the residuals, which they took as potential human influence on climate. They concluded that carbon dioxide emissions are probably only responsible for about 0.12°C of the roughly one degree C of Northern Hemisphere warming since 1881. If their assumptions are correct, the potential influence attributed to humans on recent warming can be no more than about 12 percent. The solar and human contributions to recent warming are not known, but the range of possible values is large, and the IPCC are only considering cases with a small or nonexistent solar contribution. By not considering all possibilities, they are being misleading (Connolly et al., 2021). Figure 23 compares the SORCE satellite TSI to the SILSO sunspot record.

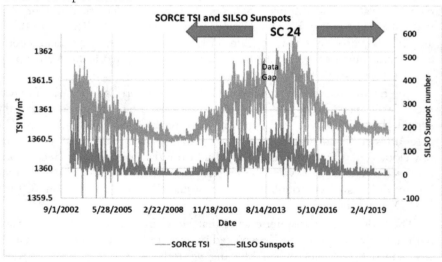

Figure 23. SORCE TSI compared to the SILSO sunspot record. Data from SORCE and SILSO.

As noted above, the more variable solar reconstructions include proxies that are sensitive to the less active portions of the Sun and are less reliant on

---

[63] (Leamon, McIntosh, & Marsh, 2020)
[64] (Soon, Connolly, & Connolly, 2015)

sunspot number.[65] Figures 23, 24, and 25 show recent TSI measurements by the SORCE TSI satellite instrument, which measured TSI continuously from 2003 until February of 2020, with one notable gap in 2013. The blue curve in the figures is the SILSO[66] sunspot number. Figure 23 is an overview of the whole record, which covers solar cycle 24. Figure 24 shows the beginning of solar cycle 24 and Figure 25 shows the end. Periods of zero sunspots still show significant solar-output variability.

Studies of stars similar to our Sun show that it is in a relatively quiet period of its life. Lockwood and colleagues write:

> "This suggests that the Sun is in an unusually steady phase compared to similar stars, which means that reconstructing the past historical brightness record, for example from sunspot records, may be more risky than has been generally thought." (Lockwood, Skiff, Baliunas, & Radick, 1992)

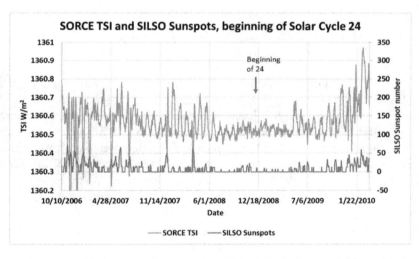

Figure 24. TSI compared to sunspot number at the beginning of solar cycle 24.

Another study, by Phillip Judge and colleagues found that extreme reconstructions, like the Hoyt and Schatten model in Figure 22, fall within the normal range for many other Sun-like stars.[67] Two specific periods of rapid solar change on our Sun are identified in red, in Figure 26, on a reconstruction prepared by Shapiro (Shapiro, et al., 2011). Both changes in solar output fall within the range observed in other Sun-like stars. The right-hand change in Figure 26, shown in red, occurs in half the stars they studied. Judge and

---

[65] (Scafetta, Willson and Lee, et al. 2019)
[66] Sunspot Index and Long-term Solar Observations (SILSO, 2020)
[67] (Judge, Egeland, & Henry, 2020)

colleagues tell us that the IPCC estimates of solar-forcing variability are an order of magnitude smaller than those observed in other Sun-like stars.

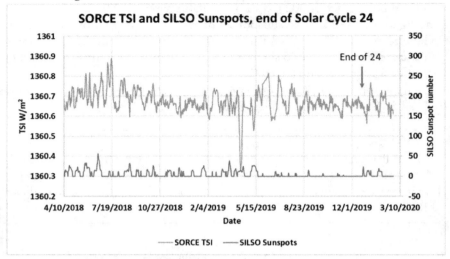

Figure 25. SORCE TSI compared to SILSO sunspot number at the end of solar cycle 24. The arrow shows the end of solar cycle 24.

Since 1750, the IPCC estimates that humans have contributed 1.1 to 3.3 W/m² of radiative forcing through greenhouse-gas emissions and the Sun has provided -0.3 to 0.1 W/m². Judge and colleagues show that, if our Sun acts like similar neighboring stars, its solar forcing on Earth's atmosphere alone could vary as much as 4.5 W/m² since 1750. After correcting this value for Earth's albedo (~30% of the radiation is reflected to space) and the fact that Earth is a sphere (divide by 4) the resulting value of 0.8 W/m² is tantalizingly close to the estimated human contribution and eight times the upper IPCC solar forcing estimate.

The Shapiro and colleagues TSI reconstruction in Figure 26 is similar to the reconstruction by Hoyt and Schatten in Figure 22. Both have a range of four to five W/m².

As explained by Ronan Connolly and an impressive list of 22 co-authors in a 2021 peer-reviewed article in *Research in Astronomy and Astrophysics*,[68] the IPCC appears to overlook, or simply ignore, the ongoing debate in how to properly reconstruct the recent TSI record and its implications for apportioning climate change between nature and humans. The IPCC have chosen one estimate, without showing the others are incorrect. Connolly and colleagues conclude that we do not know how large a role solar variability has played in recent warming and attempts to forge a consensus view either way are premature and stifle scientific progress.

---

[68] (Connolly et al., 2021)

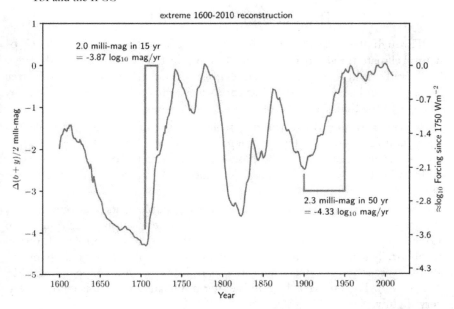

Figure 26. The solar reconstruction from Shapiro and colleagues compared to observations of other Sun-like stars. The left, 15-year period, is compatible with several stars in the study and the right, 50-year period, is compatible with over half the stars in the study. Used with Philip Judge's kind permission (Judge, Egeland, & Henry, 2020).

Connolly, et al. identified 16 long-term solar reconstructions from the peer-reviewed literature. Half of them were low variability, like the right side of Figure 21 and half were active, like the left side. All the reconstructions go back to 1700 or earlier. Then they compiled five different Northern Hemisphere temperature reconstructions, these were of rural temperatures, both urban and rural weather stations, SSTs, tree ring proxies, and glacier advance/retreat proxies. Each of the five temperature-series were statistically compared to all sixteen solar reconstructions and the residuals compared to the IPCC anthropogenic forcing time series. They found that the Sun could have contributed anything from 0% to 100% of the forcing necessary to warm the Northern Hemisphere as observed since the 19th century.

As acknowledged in the paper, their methodology is very simple and it assumes that all the time series are statistically stationary and the observations are independent, neither of which is true. It also assumes that the relationship between solar output and climate change is linear, which is also likely not true. But these are the same assumptions made by the IPCC. The Connolly, et al. study emphasizes that any proper work apportioning climate change between human and solar influences must include all appropriate peer-reviewed solar reconstructions.

The debater's views on the Sun

Through the last three chapters, we have investigated the evidence of natural climate variability in the past. We have described the evidence that very large and rapid changes in temperature occurred before the industrial revolution and were caused by variations in Earth's orbit around the Sun, by changes in the Sun itself, and by reasons unknown.

Karoly and the IPCC do not believe that natural variability plays a significant role in recent climate change, except for volcanos. In their view the Sun is nearly constant over the past 300 years, except for the 11-year solar cycle and believe the various ocean oscillations, like ENSO, cancel each other out in century or longer time scales.

Tamblyn tells us that the Sun is "surprising stable in its heat output, only showing a small variation of less than 0.05% in heat output over roughly an 11-year cycle, the sunspot cycle." (Tamblyn, 2021a, p. 4). Tamblyn emphatically states that only changes in the Sun, Earth's reflectivity (albedo), and the greenhouse effect can cause significant climate change (Tamblyn, 2021a, p. 9). Then he says the Sun is either not changing or "it is cooling very, very slightly."

Karoly tells us that the currently observed warming cannot be explained by natural climate variability, natural forcing, or changes in solar irradiance. He knows this because of climate model simulations (Karoly, 2021b, pp. 14-15). Like Tamblyn, he is confident that changes in the Sun have had no influence on recent global warming.

They believe that recent warming is controlled by changes in CO_2 and other greenhouse gases.[69] Karoly has presented evidence that this is the case and Happer has provided a rebuttal. Next, we look at the IPCC climate models. Since there are no observations that show CO_2 and other GHGs are driving climate, the models are their only "proof." Just how accurate are they?

---

[69] (Lacis, Schmidt, Rind, & Ruedy, 2010)

Chapter 5: How Accurate are the Climate Models?

Karoly separates natural forces from human influences on climate with more than 30 computer models that mostly compare well to observations of global-average-surface temperature since 1860 and points out:

"The observed significant cooling for one to two years after major volcanic eruptions—Santa Maria (1903), Agung (1963), El Chichon (1982), and Pinatubo (1991)—is simulated very well. The observed global mean temperature variations throughout the whole period lie within the range of all the model simulations with combined forcings, indicating the models simulate well the chaotic interannual variability of global mean temperature. There is very good agreement between the observed long-term global warming since the late nineteenth century and the average global warming across all the model simulations for combined natural and anthropogenic forcing." *From the Karoly statement* (Karoly, 2021b, p. 12)

He shows a figure from IPCC AR5 (IPCC, 2013, p. 879) included here as our Figure 27. It displays model results from the CMIP3 (AR4) and CMIP5 (AR5) climate models. The CMIP group uses averages of selected climate models to project future climate. The projections are used to model the impact of climate change on future societies.[70]

The effect of humans and CO₂ has not been observed

As Karoly explains, the evidence of the human impact on climate in AR5 is entirely model based (Karoly, 2021b, p. 12). The human impact on climate has never been observed or measured. Critics of the calculation, point out that the warming trends from 1910 to 1945 and 2000 to 2012 are poorly matched by models (May, 2015b). Karoly responds:

"… there is large variability in global mean temperature in the observations and the models. The observed departure in 2010 from the multi-model mean is no larger than in 1910 or in 1940 and is well within the envelope of all the model simulations." *From the Karoly statement* (Karoly, 2021b, p. 14).

---

[70] (World Climate Research Programme, 2021)

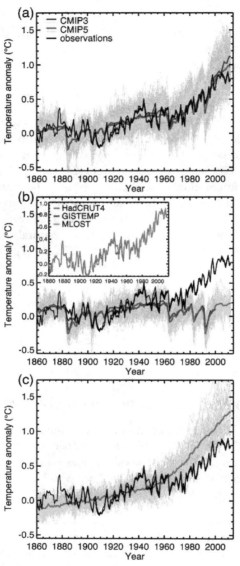

Figure 27. (a) The ensemble mean of numerous CMIP3 ana CMIP5 climate models are shown in blue and red, respectively, ana compared to global average-surface temperature observations shown in black. The range of modeled results is shown with yellow and gray shading. The minimum warming in (a) from the models can be almost a degree C lower than the maximum model value for any given date ana the error increases with time. This display models both human-causea and natural warming. (b) Same as (a) but only natural warming is simulated. (c) Same as (a) but only greenhouse-gas warming is simulated. (c) suggests that net human forcing (except for greenhouse gases) is negative, since natural solar and positive internal forcings are assumed to be zero by the IPCC. The temperature anomalies are computed relative to the mean from 1880 to 1919. Source IPCC AR5, figure 10.1 (IPCC 2013) page 879.

Thus, the poor model reproduction of warming from 1910 to 1945 and 2000 to 2012 is acknowledged, but in Karoly's opinion, the mismatch is acceptable. The model error from 1910 to 1945 ranges from -0.3° to +0.3°C, relative to observations, while total warming throughout the period is almost 0.7°C. This is nearly equal to the warming from 1975 to 2010, but human emissions of CO_2 are not considered to be significant before 1951. How interesting that both sides use the poor model match from 1910 to 1945 in their arguments, but with opposite intent.

After dismissing the significant mismatch between the model and observations from 1910 to 1945, Karoly writes: "The observed large-scale

73

increase in surface temperature across the globe since the mid-20th century is primarily due to human activity" (Karoly, 2021b, p. 4). The evidence is shown in Figure 27. Besides the model results in Figure 27, Karoly also refers to smaller scale model studies of special patterns of temperature change (Karoly, 2021b, p. 14). These local studies are often called "fingerprint detection and attribution studies" and include the studies of tropospheric and stratospheric temperature changes discussed in Chapter 4.

In response, Happer says the following:

> "I disagree. This statement is based on excessive faith in computer models. The wide availability of computers and powerful software to make color displays has been a serious problem, since it has blurred the lines between reality and virtual reality. These are not the same. In my Statement and Interview I tried to stick to real satellite pictures of visible and thermal radiation from the Earth, real measurements of ocean pH, real records of tornados, hurricanes, floods, droughts, etc. Essentially all of Dr. Karoly's claims of warming from greenhouse gases come from computer models, with lurid, threatening reds [red colors] to represent the supposedly harmful effects of the demon gas, CO_2." *Happer's detailed response* (Happer, 2021c, p. 8)

Thus, according to Happer, computer-model output is not data, it isn't even a proper scientific theory. Computer models are mathematical tools used to test an idea. They only help formalize and develop a theory; they are not the theory itself. The term "climate science" has become synonymous with "climate modeling." Unfortunately, modeling is not science, models can only be used to support a theory if they accurately predict the future. Happer doesn't believe models should be used to compute the human effect on climate until they are validated by making consistently accurate predictions. So far, there is no such model, and certainly the IPCC "ensemble mean," shown in Figure 27, is not predictive.

Karoly also mentions the atmospheric fingerprint studies already discussed, while they are compelling evidence that GHGs have some influence on global warming, they are not quantitative. Also, we see a similar fingerprint from El Niño Southern Oscillation (ENSO) events, as a careful look at Figure 18 shows. In Figure 18, notice the lower stratospheric response to the 1998 and 2016 super El Niños. We will discuss the topic of climate change attribution more in Chapter 9.

Update on Climate Models

In the last section we focused only on global average-surface temperature. Earth's surface is unstable and where most weather occurs. But climate and climate change involve more than the surface, we also need to consider the upper atmosphere.

Happer presents a comparison between CMIP5 climate model forecasted tropospheric temperatures with weather balloon and satellite measurements by Dr. John Christy in his Final Reply (Happer, 2021d, pp. 7-8). They do not match and Happer points out this falsifies the models. He continues: "In any normal branch of science, one would go back to the drawing boards, try to find out what is wrong with the models, and generate new models that are better able to predict the future." Yet, the IPCC continues to use their models to compute the human impact on climate and, in AR6, even admit that the uncertainty in the CMIP6 models is larger than it was in the AR5 CMIP5 models (IPCC, 2021, p. 7-114).

Background: Climate model accuracy

Figure 28 is a more recent graph by Dr. John Christy that shows temperature trends derived from CMIP6 models, these are the models used in the 2021 IPCC AR6 report (IPCC, 2021).

The individual models in Figure 28 are the tangled spaghetti and their mean, which the IPCC calls a "multi-model" or "ensemble" mean, is shown by the red-outlined, yellow-filled boxes. The dark green line is a weather reanalysis, and the light green line is from weather-balloon data.

The temperatures graphed are from about 10 km to 12 km altitude in the tropical troposphere. Christy supplies the interval in terms of air pressure, 300 hPa to 200 hPa.[71]

The top of the tropical troposphere is marked by the tropopause, a thin layer of nearly constant vertical temperature and an average altitude, in the tropics, of about 59,000 feet or 18 km. Thus, we are looking at temperature changes in the middle to upper troposphere.

One thing all models predict is that, if greenhouse gases are warming the surface, the middle- to upper-tropical troposphere will warm faster than the surface and will create what is sometimes called the tropical tropospheric "hot spot" (McKitrick & Christy, 2018). Figure 29 shows the hot spot generated by the Canadian climate model.

---

[71] One hPa[71] equals one millibar (mbar) which is a more common unit of atmospheric pressure. A pressure of 300 mbar occurs at about 30,000 feet (10 km) altitude and 200 mbar occurs at about 38,000 feet (12 km) altitude.

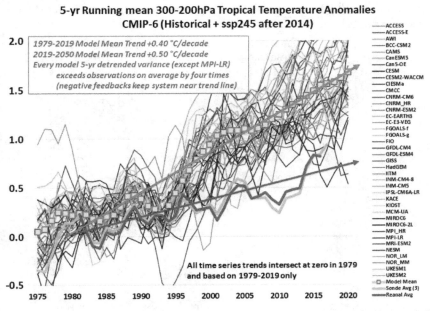

Figure 28. CMIP6 models versus weather balloon observations (light green) and weather reanalysis (dark green) in the tropical mid- to upper troposphere. Source: Dr. John Christy, used with permission.

Figure 28 is from a presentation John Christy gave to the Irish Climate Science Forum on January 22, 2021. Ross McKitrick and John Christy knew that all models predicted a hot spot in this region of the troposphere, if GHGs were increasing, so they designed this test of model validity. They applied the test to both the CMIP5 and the CMIP6 models.

In Figure 28 we see that most, if not all, of the AR6 models greatly overestimate warming in the middle- to upper-tropical troposphere. Thus, they are invalid and failed the test.

All model runs shown use historical forcing to 2014 and SSP-RCP 245, a moderate-model CO_2 emissions scenario used in IPCC AR6, thereafter. The modeled middle tropospheric warming, graphed in Figure 28, is shown in profile in Figure 29. The horizontal scale in Figure 29 is the latitude in degrees, the vertical scale is atmospheric pressure, and the colors are the warming, in degrees/decade, from 1979-2017. Red is the fastest warming, the red spot, labeled the "Hot Spot," is warming of 0.6 to 0.7°C/decade.

In Figure 18, we saw this was not the case for the global middle troposphere since 2000 as it, along with the stratosphere, declined in temperature, while the lower troposphere warmed. Figures 28 and 29 only consider the tropics; but Figure 18, which plots global temperatures, also appears to invalidate the models, and suggests that something other than greenhouse gases are causing current warming.

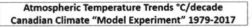

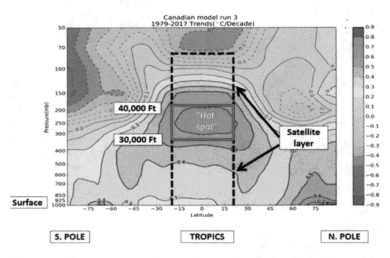

Figure 29. The tropospheric "hot spot" as seen by the Canadian Climate Model. Source (Christy, 2019).

If extra human-greenhouse-gas emissions are not included in a model run, the hot spot disappears or is greatly reduced. Further, the modelers do not tune their models to create a hot spot, like they do to match global average-surface temperature, it just appears in every run. Thus, comparing the modeled hot spot to observations is a good quantitative test of the models *and* the idea that rising CO_2 is causing global warming.

In his interview, Happer estimated the rate of lower- to middle-tropospheric warming should be about 1.2 times surface warming (Happer, 2021a, p. 18) and the concept was discussed in the debate. His reasoning was simple. If GHGs cause the surface to warm, evaporation will increase on the ocean surface. According to Happer evaporation and convection provide about half the cooling of Earth's surface because the lowermost atmosphere is nearly opaque to most infrared radiation (IR). Evaporated water carries latent heat with it as it rises through the atmosphere. Water vapor rises because it has a lower density than dry air.

As the water vapor rises through the lower atmosphere, the air expands and cools. Eventually it reaches a height where the vapor condenses to liquid water or ice (the local current cloud height). This causes it to release infrared radiation, some of the radiation warms the surrounding air and some goes to outer space. It is this release of heat that causes a temperature inversion and creates the hot spot.

Does the hot spot exist? If GHGs are warming the surface significantly, theory and all climate models say it should. Some observations (see Figure 30) show warming at 300 mbar is faster than at the surface, but it seems more related to ENSO events (El Niños) than a global CO_2 effect. Proof it is caused by increasing greenhouse gases, has been elusive.

In Figure 30, we plot the surface temperature from the European Centre for Medium-Range Weather Forecasts (ERA5) weather reanalysis versus the reanalysis temperature at 300 mbar (about 10 km altitude). The curves are for the tropics between 20°S to 20°N latitude. We tend to trust reanalysis data, after all it is created after the fact and compared to thousands of observations around the globe. As the figure shows, the surface warms at 0.15°C per decade and the atmosphere at 300 mbar warms at 0.2°C per decade.

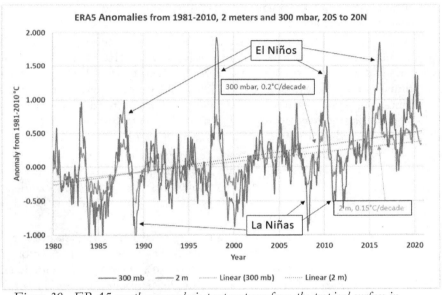

Figure 30. ERA5 weather reanalysis temperatures from the tropical surface in orange and at 300 mbar (10 km) in blue. We expect a faster rate of warming at 300 mbar in the tropics, and we see this. The El Niños are warmer at 300 mbar than the surface and the La Niñas are cooler. Source data from the World Meteorological Organization's KNMI climate explorer (World Meteorological Organization, 2021).

Both GHGs and El Niños should cause more ocean evaporation and thus more warming at 300 mbar than at the surface. Both will help form the hot spot shown in Figure 29. In Figure 30, the hot-spot-warming rate is 1.3 times that of the surface, slightly higher than Dr. Happer's estimate of 1.2.

Warming at 300 mbar is enhanced during El Niños and reversed during La Niñas. If the warming at 300 mbar reverses, and becomes cooling during La Niñas, then how much of the 300-mbar warming is due to the El Niños over the period and how much is due to GHGs? If the past 40 years had more La

78

Niñas and fewer El Niños, would the trend reverse? The answer is unclear, but it does suggest that GHGs are not controlling the warming.

In AR6, the IPCC admits that their models overestimate tropospheric warming:

> "For upper air temperature, there is *medium confidence* that most CMIP5 and CMIP6 models overestimate observed warming in the upper tropical troposphere by at least 0.1°C per decade over the period 1979 to 2014." (IPCC, 2021, pp. 3-5)

Christy's analysis in Figure 28 suggests that the models are overestimating warming from 1980 to 2014 by about 0.2°C per decade. This can be computed by comparing the green straight line with the red line. This is twice the IPCC estimate in the box above.

We can also look at the hot spot in vertical cross-section, see Figure 31 below taken from Figure 10.SM.1 of the IPCC AR5 "10SM" supplementary materials (Bindoff & Stott, 2013, pp. 10SM-6). It compares observed warming to climate-modeled warming, from 1979 to 2010, for different atmospheric pressures (left scale) and altitudes (right scale). The colored areas are the 5th to 95th percentile modeled warming for natural climate forcing (blue), GHGs only (green), and all forcings, that is anthropogenic plus natural (red).

Observations of the warming trend are shown in black and gray. HadAT2[72] is shown as a thick black line, the thin black line is RAOBCORE,[73] the darker gray band is the RICH[74] model ensemble, and the lighter gray band is a separate RICH-model ensemble.[75] The RICH-modeled dataset is based on observations, like a reanalysis.

All datasets are trimmed to the same area. The HadAT2 dataset covered the most limited area, so the others are trimmed to match. All observations fit into the modeled natural-forcings-only region of the plot (blue shaded area) below 130 mbar and all are less than the modeled greenhouse gas output for the six models used. Like the other colored regions, the blue band covers the 5th percentile to the 95th percentile of the model runs. Six models and 87 runs are represented (Bindoff & Stott, 2013, pp. 10SM-5).

In the main AR5 report we see a similar plot (Figure 32, right side) using an expanded period, from 1961 to 2010. Tropospheric temperatures of both the HadAT2 and RAOBCORE datasets are below the all-forcings and greenhouse-gas-forcings bands below 130 mbar but are above the oddly

---

[72] HadAT2: The Hadley Centre Atmospheric Temperature reanalysis dataset 2.

[73] RAOBCORE: the Radiosonde Observation Correction using Reanalysis, a weather reanalysis dataset (Radanovics, 2010).

[74] RICH: Radiosonde Innovation Composite Homogenization

[75] RICH-model: (Haimberger, Tavolato, & Sperka, 2012)

narrower natural-forcings band. The RICH datasets have moved higher and reach the lower portion of the all-forcings and greenhouse-forcings bands.

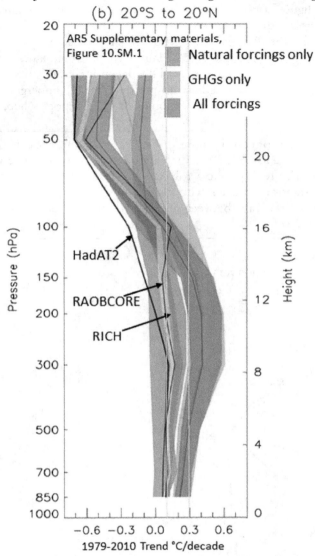

Figure 31. Cross section of the atmosphere showing temperature trends from 1979 to 2010. HadAT2, RAOBCORE, and RICH are observations, in black and grey. The red and green bars are the 5th to 95th percentile ranges of climate models that include anthropogenic forcings. The blue band is the range of the natural forcings only models. Source: after (Bindoff & Stott, 2013, pp. 10SM-6)

The reason why the blue natural-only area has narrowed in the main AR5 figure is not explained in the text or in the figure caption. It must be artificial since this figure covers 18 more years (1961–1978) than Figure 31, which should increase the model spread, not narrow it. The AR5 main report suggests that the RAOBCORE and RICH radiosonde observations are questionable and reference Leopold Haimberger and colleagues, who explicitly discusses the significant differences in the data prior to and after 1979, and the necessity to make a serious adjustment in that year (Haimberger, Tavolato, & Sperka, 2012). Yet, in the main section of the AR5 report, where our Figures 31 and 32 (right side) are compared, they say:

> "An analysis of contributions of natural and anthropogenic forcings to more recent trends from 1979 to 2010 (Supplementary Material, Figure S.A.1 [our Figure 31]) is less robust because of increased uncertainty in observed trends (consistent with (Seidel, et al., 2012)) as well as decreased capability to separate between individual forcings ensembles." (IPCC, 2013, p. 892)

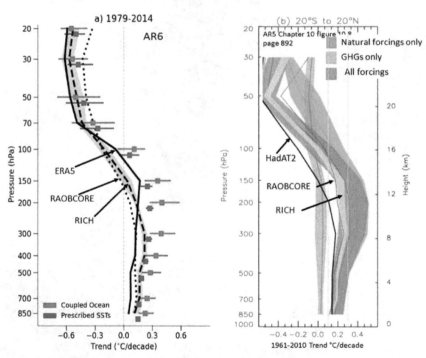

Figure 32. A comparison of tropical (20S to 20N) vertical profiles from AR6 and AR5. The black and gray lines are observations, and the model output is shown in red, blue, and green. Sources: (IPCC, 2021, pp. 3-162), (IPCC, 2013, p. 892).

The quote above suggests that the observational uncertainty of the radiosonde data throughout the period from 1961 to 2010 is lower than the uncertainty from 1979 to 2010; that is, Figure 32 (right side) is more accurate than Figure 31, and they refer to (Seidel, et al., 2012) as evidence. Dian Seidel and colleagues studied the lower atmosphere, below 2.5 kilometers with radiosonde data. They did not study the data we are focused on between eight and twelve km.

Seidel and colleagues study many sources of uncertainty in their calculations of the atmospheric-boundary-layer height, which is the top of the lowermost layer of the atmosphere (often below cloud base level) where most weather occurs. However, their paper makes no mention of a study of the uncertainty in radiosonde-measured temperature. The principal uncertainties they mention are wind speed, their interpolations of the vertical profile, and the vertical resolution of the data.

On the other hand, Haimberger and colleagues explicitly mention that the main uncertainty is the splicing together of ERA-40 pre-satellite and ERA-Interim post-satellite (1979 and later) radiosonde data. The 1979 transition adds significant uncertainty to the result, and inclusion of the earlier data does not increase the accuracy of the calculations, as suggested in AR5, it decreases it. Haimberger and colleagues write:

> "The properties of the forecasts are quite different between ERA-40 and ERA-Interim. Most notably, ERA-40 uses 6-hourly cycling and a three-dimensional variational data assimilation (3D-VAR) system, whereas ERA-Interim uses a considerably more advanced 4D-VAR assimilation system with a better forecast model, 12-h cycling, and variational bias correction of satellite data. No bias correction has been applied to radiosonde temperatures in ERA-40 during the period 1958–79. In ERA-Interim, the radiosondes have been adjusted to remove both the annual mean bias, using RAOBCORE v1.3 adjustments" (Haimberger, Tavolato, & Sperka, 2012)

Thus, the assertion in AR5 that Figure 32 (right side) is more accurate than Figure 31 is not supported by their sources. However, both figures suggest that the warming computed by the models is too fast.

The left atmospheric profile in Figure 32 is from AR6, Figure 3.10, page 3-162. The red lines and boxes show the mean and range of 60 CMIP6 models. The blue lines and boxes show the mean and range of 46 CMIP6 models that were run with fixed SSTs. The blue boxes and lines are meant to show that part of the error in the red boxes and lines is due to the models overestimating SST. The remaining error is due to overestimating air temperature.

The plot illustrates that the AR6 models are overestimating the rate of warming in the tropical middle troposphere by a larger margin than the AR5 models. The highest error has increased from a peak of about 0.5°C/decade in AR5 to about 0.6°C/decade in AR6. The weather balloon radiosonde observations[76] cover different time periods and areas in AR5 and AR6, but otherwise are similar. The weather reanalysis datasets are different. In the AR5 plot the Hadley Centre reanalysis dataset is used and in the AR6 plot the ERA5 dataset is used, but, again, they are similar. The observations through the critical interval from 150 to 300 mbar are about zero to 0.2°C/decade. The AR6 modeled temperatures have simply moved much higher than the models in the already high AR5 report.

AR6 acknowledges that the models are overestimating tropical tropospheric temperatures but cannot explain the difference and they do not admit that this difference, which covers half of Earth's surface, invalidates their models. They stick by their conclusion "that anthropogenic forcing, dominated by greenhouse gases, was the main driver" of tropospheric warming.

They do acknowledge that many studies suggest that the high AR6 estimates of climate sensitivity are to blame for the overestimate of tropospheric warming.[77] In AR6, unlike in previous reports, they estimate a range of ECS independently of the models.[78] The model results are then expected to produce an ECS in this predetermined range.

In AR6, the IPCC admits that "despite decades of model development, increases in model resolution and advances in parametrization schemes, there has been no systematic convergence in model estimates of ECS. In fact, the overall inter model spread in ECS for CMIP6 is larger than for CMIP5 …"[79] For this reason, AR6 determined ECS and TCR independently of the general circulation models and used it to flag models that lie outside the approved range.[80]

Regardless of how far the observations are below the model projections, they are still below them, and the models are clearly exaggerating the hot spot. The erroneous hot spot disappears if GHG forcing is reduced to zero. Some excess warming of the 200 to 300 mbar region of the tropical troposphere may be taking place, and may be due to GHG forcing, but the warming, and the forcing are much less than calculated by the IPCC models. How much of the extra observed tropospheric warming is due to GHGs and how much is due to ENSO events (see Figure 30) is unknown.

Comparing Figures 31 and 32 and the text about these illustrations in AR5 and AR6 suggests the IPCC is trying to dismiss this clear evidence that their

---

[76] (RICH 1.5 and RAOBCORE 1.5)
[77] (IPCC, 2021, pp. 3-24 & 3-25)
[78] (IPCC, 2021, p. 7-92)
[79] (IPCC, 2021, p. 7-113, 114)
[80] (IPCC, 2021, p. 7-115)

models are not working. This does not inspire confidence in their climate-science reporting.

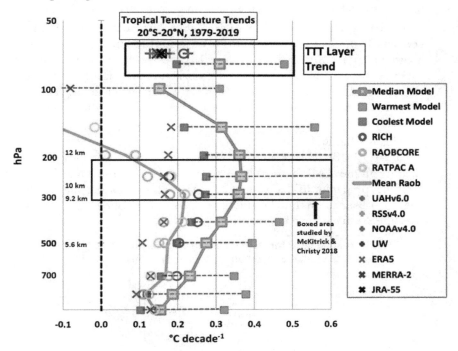

Figure 33. A plot of model tropospheric temperature trends versus observations. Model medians and ranges are in dark blue and red on the right. Observations are in yellow, green, purple, and light blue on the left. In the critical "hot spot" range of 200 to 300 hPa, studied by McKitrick and Christy, there is no overlap between the models and the observations. The plot is from the BAMS State of the Climate Report for 2019 (Blunden & Arndt, State of the Climate in 2019, 2020).

Figure 33 is another example of the mismatch between observations and climate models in this critical part of the atmosphere. It is from the *BAMS* (*Bulletin of the American Meteorological Society*) *State of the Climate Report* in 2019 (Blunden & Arndt, 2020). There is no overlap between the models and the observations between 200 and 300 mbar. Further, some of the model values are off scale to the right; exceeding 0.6°C/decade.

Happer concluded that the IPCC climate model's failure to get the tropical middle troposphere warming rate right falsifies them. We agree. Christy discovered this problem well before the CMIP5 model results were published in AR5 in 2013 and reported it to the IPCC. In fact, he reported that the observations in the tropical mid-troposphere agreed with the models best when GHG forcing was not included in the model.

They chose to ignore the model mismatch then, and now the problem is worse, and they are still ignoring it. The models are clearly invalid and cannot be used to compute the human impact on climate change.

Global-surface-temperature datasets, like Met Hadley Centre HadCRUT versions 4 and 5, are measurements used to tune the models. Thus, as Karoly mentions in his Interview on page 13, the modelers adjust their model parameters to match the observed global average-surface temperatures as closely as possible. Standard practice, when evaluating any computer model, is to compare model results to critical observations that the models are not explicitly tuned to.

This is what Ross McKitrick and John Christy do in Figure 34 (McKitrick & Christy, 2018). This is a more rigorous and formal statistical study of the validity of the CMIP5 models, than we have shown above. Figure 34 is carefully prepared to highlight the differences between the models and relevant observations in the middle-tropical troposphere. The atmospheric layer they study is boxed in Figure 33.

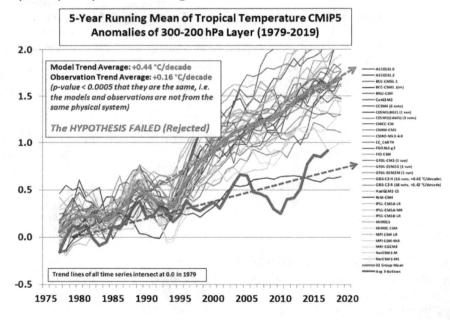

Figure 34. *CMIP5 models versus weather balloon observations in green in the tropical middle to upper troposphere. Source: Dr. John Christy, from his Irish Climate Science Forum talk in 2021.*

We have seen that the AR6 report acknowledges that climate model uncertainty has increased from AR5 to AR6, on page 4-20. Observations are shown in green in both Figures 34 and 28. Are the differences between model predictions and observations in AR5 statistically significant?

It turns out that nearly every AR5 model fails this comparison with observations at a statistically significant level. The details of why most of the ensemble models fail can be seen in McKitrick and Christy's paper.[81] All model runs shown in Figure 34, use historical forcings until 2006 and AR5 emissions scenario RCP 4.5 thereafter. The spread in results is huge, some go off scale in 2010. This is not a dataset one should average. Averaging multiple runs from one model makes some sense. Averaging runs from different models, with different assumptions and structures makes no sense. They are incompatible and the resulting average is meaningless.

The "ensemble mean"

Neither Dr. Happer nor Dr. Karoly discuss the IPCC procedure of averaging the results from multiple models into one so-called "ensemble" or "multi-model" mean. However, as mentioned above, to make his case that humans are causing most of the recent global warming, Karoly uses the "average of all the different climate-model simulations" (Karoly, 2021b, pp. 12-14).

Background: Averaging climate model results

Choosing one model, that matches observations best, is normal best practice. We have not seen a good explanation for why CMIP5 and CMIP6 produce ensemble means. It seems to be a political solution to a scientific problem. This is addressed in AR6 in Chapter 1,[82] where they refer to averaging multiple models, without considering their accuracy, as "model democracy." It's unclear if they are being sarcastic.

AR6 discusses weighting the models according to their performance and their dependance upon other models, since many models share code and logic, but could not find a robust method for determining the weights. In the end, they classified the models based first on observations prior to 2014 and second on their modeled ECS and TCR,[83] as discussed in AR6, Chapters 1 and 4.[84] These latter two values, as computed by the ensemble mean and ensemble members, were compared to ECS and TCR values determined independently of the models. The AR6 modeling process resulted in higher projected warming than in AR5. In Chapter 4 they admit that much of the increase was due to the higher ECS and TCR values used in the AR6 assessment. It appears that the AR6 "independently" assessed ECS and TCR caused the models to move farther from observations than the already high values in AR5.[85]

---

[81] See their 2018 *Earth and Space Science* paper (McKitrick & Christy, 2018)

[82] AR6, page 1-96

[83] Transient Climate Response

[84] AR6 pages 1-96, 1-97, 4-22 to 4-23, and 4-4.

[85] (AR6, 4-25)

The IPCC, in AR4, AR5, and AR6, often conflate models and the real world, so constraining their model results to an independently predetermined range of climate sensitivity is especially worrisome. Models are one of the sources for climate sensitivity because both ECS and TCR are model-based values, so their technique is partially circular. Further, models are used to forecast future temperatures.

One reason they give in both AR5 and AR6 for using an ensemble mean is they think it separates "natural variability" (or noise) from model uncertainty and trends so they can observe longer-term modeled climatic changes more clearly. In AR6, they acknowledge it is difficult to separate natural variability from model uncertainty. They tried separating them by duration, that is, by assuming that short-term changes are natural variability and longer-term changes are model uncertainty. But they found that some natural variability is multi-decadal (AR6 4-19). Internal natural variability via ocean oscillations, such as the AMO[86] or the PDO,[87] have a long-term (>60 years) effect on global and regional climate (Wyatt & Curry, 2014).

Conflating natural variability with short-term noise is a mistake, as is assuming natural variability is short term. Ocean oscillations are not noise, they are a long-term component of natural variability. The difficulty in separating natural variability from model uncertainty is discussed in AR6, Chapter 4.[88] It is not clear that the CMIP6 model uncertainty is properly understood.

The CMIP models have a tough time simulating the AMO and PDO. They might have features that approximate these natural oscillations in time and magnitude, but they are out of phase with observed temperature records and each other. A careful look at the projected portions of Figures 28 (post 2014) and 34 (post 2005) will confirm this timing problem. Thus, when the model output is averaged into a multi-model mean, natural ocean oscillations are probably "averaged" out.

The model results shown in Figures 34 and 28 resemble a plate of spaghetti. Natural climate variability is cyclical,[89] so this odd practice of averaging multiple models erroneously makes it appear nature plays a small role in climate. Once you average out nature, you manufacture a large climate sensitivity to CO_2, and erroneously assign nearly all observed warming to human activities.

The IPCC included many models in their AR5 ensemble that they admit are inferior. Some of the models failed a residual test, indicating a poor fit with observations. They also included models that did not detect a greenhouse-gas

---

[86] Atlantic Multi-decadal Oscillation
[87] Pacific Decadal Oscillation
[88] AR6, pages 4-18 to 4-19.
[89] (Wyatt & Curry, 2014), (Scafetta, 2021), and (Scafetta, 2013)

response.[90] The inclusion of models with a poor fit to observations corrupts the ensemble mean.

It seems that they are attempting to do "consensus science" and, for political reasons, are including results from as many models as possible. This is an admission they have no idea how climate works. As Michael Crichton famously said on January 17, 2003, at the California Institute of Technology:

> "I regard consensus science as an extremely pernicious development that ought to be stopped cold in its tracks. Historically, the claim of consensus has been the first refuge of scoundrels; it is a way to avoid debate by claiming that the matter is already settled."

In Professor Happer's words:

> "A single, incisive experiment is sufficient to falsify a theory, even if the theory accurately explains many other experiments. Climate models have been falsified because they have predicted much more warming than has been observed. ... Other failures include the absence of the predicted hot spot in the upper troposphere of tropical latitudes." (Happer, 2021d, p. 6)

---

[90] (IPCC, 2013, p. 882)

Chapter 6: How accurate are our temperature measurements?

> **TheBestSchools:**
>
> "There are a number of inherent difficulties in obtaining reliable data about the past behavior of the atmosphere. Perhaps the most important of these is the weakness of the 'signal' (the secular change in temperature measured in tenths of a degree Celsius), which is some three orders of magnitude smaller than the range of the geographic temperature variation, not to mention the normal diurnal and annual temperature variability, over the surface of our planet. ... ought we to place so much confidence in our models, which after all are only as good as the empirical data we feed into them?" The Karoly Interview (Karoly, 2021a, p. 13)

Karoly answers:

> "This question suggests that the climate models are based on empirical data, like economic models. That is wrong. ... Observational data is key to evaluating the performance of the models, but it is not fed into them" (Karoly, 2021a, p. 13)

Climate models are tuned to observed global warming

As Karoly describes, the global surface-temperature data are not fed into climate models, but as discussed above the model results are "tuned" to global surface-temperature observations. This is done with adjustable model parameters that *are* fed into the models.

Karoly continues:

> "We have to consider the forced response (the signal) to the internal variability (the noise): the variations in space and time." (Karoly, 2021a, p. 14)

Thus, Karoly is dividing each weather measurement, over time, into a forced response and noise. He equates noise with internal variability, as discussed in the previous chapter. Since he calls it noise, he is assuming that internal climate variability is random and not ordered; a presumption disputed by many researchers.[91]

The thrust of TheBestSchools question has less to do with how the models are adjusted to match estimated global average-temperatures and more to do

---

[91] Such as Marcia Wyatt, Christopher Moy, and Nicola Scafetta (Wyatt & Curry, 2014), (Moy, Seltzer, & Rodbell, 2002), and (Scafetta, 2013).

with the accuracy of the record. Over 70% of Earth's surface is water, yet our estimates of ocean-surface temperature are less accurate than the estimates on land and mixing the two to obtain a global average is difficult (Morice, et al., 2021).

Merging land and sea surface temperatures

As Happer mentions in his Interview on page 19, land surface-temperature measurements are heavily influenced by rapid urbanization and the resulting "urban heat-island" (UHI) effect. Temperatures are warmer in cities than in the surrounding countryside, so as rural weather stations are surrounded by growing cities, they show an artificial warming trend (Scafetta, 2021). Another serious problem is that many weather stations have been poorly maintained and are poorly located (Fall, et al., 2011). Further, weather stations are widely scattered around the world and, until recently, were only located on land.

Karoly does not believe that the urban heat island effect is a problem (Karoly, 2021b, p. 6). He believes adjustments made to the data compensate well for it and other issues. The principal adjustment to the land and ocean measurements are made by "homogenization" algorithms. They compare each weather measurement to its neighbors and smooth through anomalous values (Menne & Williams, 2009a). The problem with homogenization of the urban heat island effect, is that it smears the extra urban warming over large areas. The UHI effect is due to human influence but is unrelated to greenhouse gases. As Happer writes, it is due to replacing transpiring trees and grass with concrete and other heat-generating urban features, like cars and air conditioners.

Karoly notes that the uncertainty in the global average temperature estimates is about ±0.1° in the twentieth century and ±0.2°C in the 19th century, which is in line with other estimates (Karoly, 2021b, p. 6). Comparisons with satellite data and comparing land to ocean warming suggests these ranges may be optimistic.

Background: Surface temperature estimates

The UHI effect and weather-station maintenance are issues on land; but since 70% of Earth is covered with water the global average-surface temperature is mainly driven by SST estimates. John Kennedy studies SSTs for the Met Hadley Centre at the University of East Anglia in the U.K. and maintains its global HadSST dataset.[92] He describes the errors in his SST estimate in a pair of *Journal of Geophysical Research* papers.[93]

---

[92] (Kennedy J. , Rayner, Atkinson, & Killick, 2019)

[93] "Reassessing biases and other uncertainties in sea surface temperature observations measured in situ since 1850" parts 1 and 2 (Kennedy J. J., Rayner, Smith, Parker, & Saunby, 2011), (Kennedy J. J., Rayner, Smith, Parker, & Saunby, 2011b).

Unlike on land, SST measurements are historically from ships. The ships are constantly moving, and prior to 1940, the temperature measurements were made in buckets of water dropped over the side. As the buckets were hauled back up to the deck, they lost heat, then sometimes were taken inside a cabin before the temperature was measured. The buckets also varied in size and composition.

The Hadley Centre has divided the world into latitude and longitude grid cells. Each cell that has sufficient measurements over a given month, is averaged after appropriate corrections are made. Ships move from grid to grid with time and sometimes they use the same sampling procedure and sometimes different. Records are sometimes complete and sometimes not.

More recently, since World War II, ships have been fitted with thermistor thermometers on their water-intake ports, which measure the temperature of the water used for cooling purposes as it is brought into the engine room. The thermometers are not always optimally placed and can be affected by the warm engines on the other side of the hull. Further, a ship's draft can affect the depth of its water intake significantly.

The most recent change occurred between about 1990 and 2005 when fixed-location and drifting buoys were deployed throughout the oceans. These devices measure temperature and other ocean characteristics very accurately at known depths. ARGO floats measure parameters at depths as great as 2,000 meters, but all measure surface temperature at or near the SST reference depth of 20 cm. There are thousands of these buoys around the world and since 2005 we have a pretty accurate picture of the global average SST. Before 2005, regardless of Kennedy's raw data corrections, the picture is speculative.

The step changes in measurement quality at about 1945 and 2005 also must be dealt with. As Kennedy and colleagues state, "The uncertainty of global average SST is largest in the early record and immediately following the Second World War." As illustrated in the upper graph in Figure 35, the measurement accuracy is poor prior to World War II, but the bias uncertainty is less because all the measurements were made in roughly the same way. For 1940s and 1950s data, the measurement uncertainty is less, but the bias uncertainty is high because it is often unclear how the measurements were made. Kennedy and colleagues write that their corrected SST data are extremely close to the raw buoy data after 1997 and suggest the buoy data could be used alone and uncorrected after that. Others place the quality cutoff in 2005, but either way, the data, and trends prior to 1997 are suspect.

Kennedy and colleagues estimate the uncertainty in the global average SST to be about ±0.12° in the mid-1800s, and ±0.08° in 1950. It then drops rapidly to around ±0.055° by 1979. These uncertainty values are lower than those given by Karoly, but in the same ballpark. Since 2005, it has stayed under ±0.055°C most of the time. This is illustrated in Figure 35. Kennedy and colleagues acknowledge that these uncertainty estimates are not comprehensive, they are simply a first estimate. The actual uncertainty in global SST may be higher, as Karoly suggests.

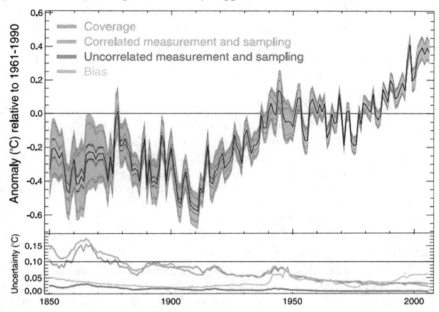

Figure 35. The lower graph is the global average SST anomaly uncertainty, as estimated by Kennedy. The light blue line is the total measurement and coverage uncertainty. The light magenta line is bias uncertainty. The shading in the upper graph shows the switch from measurement uncertainty to bias uncertainty after WWII. Total uncertainty reaches a minimum of about ±0.06°C after 2000. (Kennedy J. J., Rayner, Smith, Parker, & Saunby, 2011b)

HadSST version 4 suggests that the global ocean surface has warmed about 0.5°C since 1950, taking the uncertainty computed by Kennedy into account this could be as low as 0.2°C. This can be seen in Figure 35 by comparing the highest value in 1950 to the lowest value in 2010. There is one final issue in computing a global average-surface temperature. SST is measured at a depth of 20 cm and land-based surface air temperatures (LSAT) are measured at an altitude of two meters.

Background: AR6: GSAT versus GMST

When the IPCC models compute the global average-surface air temperature (GSAT), they compute it at two meters altitude (above the ground or ocean surface). The GSAT value has no organized "entirely observational dataset" to compare to the models. The databases for global mean surface temperatures (called GMST) use LSAT combined with SST, yet the air temperature two meters above the ocean surface is physically distinct from SST (IPCC, 2021, p. 2-37).

In Chapter 2 of the IPCC AR6 report, they discuss the difference between the air temperature at two meters above the sea surface and the SST at a nominal depth of 20 cm. In their final publicly released draft report they conclude that "long-term changes in GMST and GSAT are presently assessed to be identical." (IPCC, 2021, p. 2-6). This was also the position of all previous IPCC reports.

Previous drafts of AR6 claimed that GSAT was warming faster than GMST, and all CMIP6 models agree on this point. Yet, observations go both ways. Sometimes GSAT warms faster than GMST, and sometimes it is the other way around. The earlier drafts computed a model-based and weather-reanalysis correction of four percent. That is, they suggested that GMST's rate of warming should be increased four percent and assigned to GSAT. This was all removed from the final draft, which used GMST observed warming and assumed that equaled GSAT long term. Obviously, the AR6 authors did not agree on this, but the final decision to ignore the differences between the two datasets and return to GMST was made.

It seems odd to devote so much space in AR6 to the difference between GSAT and GMST after deciding to ignore it. Further, it seems trivial compared to the difference between model projections and observations in the tropical troposphere.

Surface Measurements versus Satellite

Chapter 2 of AR6, asserts that "Global Mean Surface Temperature (GMST) is a key indicator of the changing state of the climate system." (IPCC, 2021, p. 2-10). While true, it is not the only way to measure global warming. The atmosphere extends more than 84 km (52 miles) above the surface and the average ocean depth is 3,688 meters (12,100 feet). The oceans store over 99.9% of the "heat" (technically the thermal energy) contained in the entire atmosphere/ocean system,[94] and the lower two meters of the atmosphere only contains 0.00002% of the thermal energy. The surface is chaotic, and where weather events occur, but it holds an insignificant amount of the energy driving Earth's climate (May, 2020e).

Satellites can uniformly measure the average temperature of a larger thickness of the lower atmosphere. As Happer explains in his Interview on pages 27-28, microwave radiation emitted from oxygen atoms in the atmosphere can be used to estimate a "brightness" temperature, which can then be converted to an actual atmospheric temperature. No correlation to any other measured temperature is required. The actual temperature of the atmosphere is calculated by comparing the measurement to the brightness

---

[94] (Schmitt, 2018)

temperature of deep space (2.7K) and to a target of known temperature within the satellite (Spencer & Christy, 1990).

On this topic, there was another opinion. Tamblyn asserts:

> "Stitching together raw data from multiple satellites is very complex. Thus, the satellite datasets are much less accurate than the surface temperature datasets.
>
> "Professor Happer's stronger emphasis on satellite temperature measurements does not agree with the experts on the subject." (Tamblyn, 2021b, pp. 7-8)

Due to interference and clouds, this technique does not work close to the surface, so satellite atmospheric temperatures cannot be directly compared to surface measurements. The satellite measurements are best for measuring air temperatures in the middle and upper troposphere and the lower stratosphere.

Background: Satellite accuracy

Above, we mentioned that the Hadley Centre's best current estimate of global monthly average-SST accuracy was ±0.055°C and Karoly's was ±0.1°C. That is not quite as good as Spencer and Christy's monthly average accuracy estimate from satellite data of ±0.011°C (Spencer & Christy, 1990).

Weather-balloon (radiosonde) datasets estimate temperatures high enough in the atmosphere to be compared directly with satellite estimates. Christy, Spencer, and Braswell compared the two in a 2000 *Journal of Atmospheric and Oceanic Technology* paper[95] and estimated the error to be about 0.013°C. This is very similar to the value from Spencer and Christy, 1990. After carefully comparing the two datasets, they found that if they assumed a total error of ±0.06°C, similar to the HadSST dataset, the radiosonde and satellite data agreed. A portion of the ±0.06°C error is in the radiosonde dataset, which is not perfect.

Given this analysis, we can comfortably assume that the satellite data is at least as accurate as the surface data, if not more so. Besides being as accurate, the satellite data covers a larger volume of the atmosphere, and uniformly covers much more of the globe than the surface data.

Tamblyn seems to think that, because the satellite-temperature record has been created using data from multiple satellites, it necessarily means it is less accurate (Tamblyn, 2021b, pp. 7-8). That is incorrect. The data are merged with an accurate procedure described in (Christy, Spencer, & Braswell, 2000).

---

[95] "MSU Tropospheric Temperatures: Dataset Construction and Radiosonde Comparisons" (Christy, Spencer, & Braswell, 2000)

Happer's Statement contains one of Christy's plots, as Figure 13 (Happer, 2021b, p. 30). It is similar to our Figure 34. Here is what Happer has to say about satellite versus ground measured air temperatures:

> "In contrast to ground stations, which originally used mercury or alcohol thermometers to measure surface temperatures or sea water temperatures at sparse sample locations, satellites measure temperature remotely from orbit over most of the globe by recording the intensity of thermal upwelling radiation at frequencies close to 60 GHz, and with wavelengths of about five mm. ... The 'brightness' at frequencies close to 60 GHz can be measured very precisely with the aid of calibrated blackbodies aboard the satellite. ..." (Happer, 2021a, pp. 26-27)

We have already described various problems with the surface datasets. We also need to remember that surface measurements are obtained in a zone where there is great volatility and a large diurnal variation. Satellites measure higher in the atmosphere, in a more stable environment, thus are probably better suited to estimating changes in climate, as opposed to measuring changes in weather.

In the previous chapter, we discussed the tropical tropospheric hot spot between 200 and 300 mbar (an altitude of about 9 to 12 km). Happer shares his expert opinion that the middle troposphere should be warming more and faster than the surface as CO_2 rises, in his interview (Happer, 2021a, p. 18). The tropical weather balloon and weather reanalysis data in Figure 33 shows this is true in the tropics, although, as we have seen, the climate models overestimate the difference. The satellite data in Figure 18 shows the situation is complex since the global trend—from 2000 to today—in the middle troposphere, is cooling as the lower troposphere is warming. In Figure 36, we see that the global lower troposphere warms at the same rate as the SSTs, but the HadCRUT5 global land-plus-ocean surface warming trend is much higher than the lower troposphere. Observed trends show the real world is more complex than our models.

The global surface and lower troposphere trends match—at least over the oceans. And the HadCRUT5, land-plus-ocean global temperatures warm much faster than the lower troposphere, the opposite of what Happer and the IPCC global climate models predict if CO_2 and other GHGs are a major factor in global warming. It is also possible that the HadCRUT land-only record is overestimating global warming due to the UHI effect or some other error.

Tamblyn tries to explain this discrepancy by asserting that satellite measurements are less accurate than surface-temperature measurements, which is contrary to what the respective dataset custodians have computed and published. If the satellite and surface data since 1979 are all reasonably accurate, then either the data show the situation is more complicated than we currently think, or CO_2 and the other GHGs have only a minor, and undetectable, effect on global warming.

As the plots in the last chapter make clear, the differences between modeled temperatures and observations are clearest in the tropics between 200 mbar and 300 mbar, but a difference also exists as low as 700 mbar or about three kilometers and below (see Figure 33).

Happer, clearly states the satellites are not measuring surface temperature and have consistent, nearly global coverage. Ground measurements, on the other hand, are sparse, irregularly spaced, and made with many different devices.[96]

Background: Comparing HadCRUT to UAH

Figure 36 shows the latest global temperature anomaly trends from Kennedy's HadSST4, the HadCRUT5 land-plus-SST, and the UAH version 6 lower-troposphere-land-plus-ocean datasets.

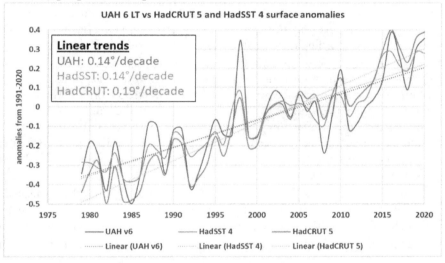

Figure 36. HadSST4 is in gray, HadCRUT5 global land-plus-SST in orange, and the UAH global land-plus-ocean in blue are compared. HadSST4 and UAH have nearly the same warming trend of 0.14°C/decade. HadCRUT land-plus-ocean is higher at 0.19°C/decade.

Figure 36 shows the HadSST4 and UAH lower troposphere warming trends are nearly identical at 0.14°C/decade. The major difference between the two curves results from the ENSO events. Temperature variations caused by El Niños and La Niñas are more extreme in the lower troposphere than on

---

[96] (Kennedy, Rayner, Atkinson, & Killick, 2019) (Karl, Williams, Young, & Wendland, 1986).

the surface and are represented by greater fluctuations on the UAH curve. This is expected, since—as previously explained—these events greatly affect tropical Pacific evaporation. The El Niños increase vapor pressure by warming the surface water and the La Niñas do the opposite when they cool the ocean surface in the tropics. Two prominent El Niños are visible in 1998 and 2016. The lower troposphere warming, during El Niños and cooling during La Niñas, are slightly delayed as one would expect. Two strong La Niñas are visible in 2008 and 2010–2011.

The HadCRUT5 dataset, which is a combination of the CRUTEM5 land-surface air temperatures and HadSST4 sea surface temperatures, plots well below the other two early in the period and above toward the end. The 36% increase in warming rate is mostly due to adding the CRUTEM5 land-measured surface temperatures to the SSTs, even though land is only 29% of Earth's surface. HadCRUT5 also adds Arctic kriging (map-based extrapolation and interpolation), which attempts to capture recent Arctic winter warming. We also need to recognize that UAH does not have good coverage over the poles, but does Arctic warming really make that much difference in the warming trend? Probably not. Antarctica is not a factor as it has warmed very little, if at all, according to the data at climate4you.com.

Figure 36 suggests three important conclusions. First, El Niño years support the idea that a warmer surface will lead to more evaporation and the additional water vapor will carry latent heat into the troposphere. As the vapor condenses to droplets, usually at the local cloud level, the latent heat is released which warms the surrounding air. The lower troposphere UAH temperature, plotted in Figure 36, is most sensitive to temperatures in the 600 to 700 mbar (~3 to 4 km) region, but it includes some oxygen brightness emissions below 2 km, which is a common altitude of lower clouds.

Second, since the overall rate of HadSST warming is about the same as the UAH lower troposphere warming, it suggests that the extra warming in the HadCRUT5 land-plus-ocean dataset is suspect, Arctic winter warming or not.

Third, if HadCRUT5 is correct, it means the surface is warming faster than the lower and middle troposphere. If this is true, the climate models and Happer suggest that the warming is *not* due to greenhouse gases. It could be that additional warming in the troposphere, above two or three kilometers, is due to El Niños and not due to GHGs. It is hard to accept that all the data plotted in Figure 36 is both correct *and* consistent with the idea that greenhouse gases are causing most of our surface warming.

The debate over whether the world surface is warming 0.19°C per decade (HadCRUT5 surface) or 0.14°C per decade (HadSST and UAH lower troposphere) is a bit beside the point. As originally noted by the late Stan Grotch and quoted by Lindzen and Christy (Lindzen & Christy, 2020), this warming is a small residue of a wide scatter of temperature measurements. The weather stations and buoys observe a wide range of temperatures every day, every month, and every year. The seasonal range of station-temperature anomalies seen is over 16°C, as shown in Figure 37. The global warming, as reported by the IPCC and NOAA, is tiny compared to ordinary daily temperature variations.

We will recall from Chapter 2, Figure 10, that according to the HadCRUT5 land-plus-SST records, the Northern Hemisphere has warmed much faster than the Southern Hemisphere in recent years. HadCRUT5 shows the Northern Hemisphere warming 0.28°C per decade and the Southern Hemisphere warming at only 0.11°C per decade.

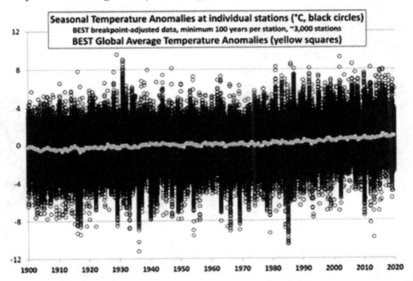

Figure 37. Seasonal temperature anomalies observed by BEST (Berkeley Earth Surface Temperatures). The global average anomaly is shown with yellow boxes. Source: (Lindzen & Christy, 2020).

The UAH records show a much smaller difference between the hemispheres. Figure 38 shows only the Northern and Southern Hemisphere curves from Figure 9. Using the UAH Lower Troposphere records the difference between the two hemispheres is only 0.05°C per decade.

The HadCRUT5 Northern Hemisphere warming rate is 0.28°C per decade and the UAH rate is 0.16°C per decade. This sort of large difference between the surface and lower troposphere is questionable, especially since the Southern Hemisphere surface and lower troposphere rates match. Comparing Figures 9 and 10, we see that the UAH Southern Hemisphere warming rate and the HadCRUT5 Southern Hemisphere surface warming rate are both 0.11°C per decade. Something is probably wrong in one of the datasets, and it is probably in the HadCRUT5 dataset since it is the outlier. We can be fairly certain of this, since the UAH global warming rate matches the HadSST warming rate in Figure 36. But, if we are wrong, and all the data are accurate, the surface is warming faster than the lower troposphere, which means, according to Happer and the IPCC, that increasing GHGs are not responsible for most of the warming. The consensus can't have it both ways.

We can see, from Figure 10 in Chapter 2, that according to the Hadley Climatic Research Unit the Northern Hemisphere has warmed 0.7°C more than the Southern Hemisphere since 1979. The satellite-temperature-lower-troposphere record is directionally the same but the difference in hemispheric warming rates is much less, as shown in Figure 38. Specifically, the UAH global satellite-warming difference is about 0.2°C, a full 0.5°C less than the HadCRUT5 difference. Clearly, the HadCRUT5 temperatures for the Northern Hemisphere are suspect.

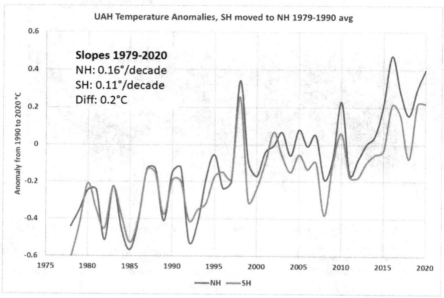

Figure 38. The UAH satellite lower troposphere temperature records for the Northern and Southern Hemispheres. The difference is much less than the HadCRUT 5 difference shown in Figure 10. Data source: Dr. Roy Spencer.

A difference in hemispheric warming rates is characteristic of solar and orbital forcing due to the tilt of Earth's axis relative to the orbital plane. It is not characteristic of forcing due to a well-mixed greenhouse gas like CO_2. The HadCRUT5 record is suspicious and so is the difference between it and HadSST in Figure 36. Perhaps the urban heat-island effect, combined with the Hadley Centre homogenization algorithm has contaminated the land-only HadCRUT5 record and distorted it. Further, the surface warming faster than the lower troposphere is strange, it should be the other way around.

People have adapted to, and live in, places with temperatures that range from −40°C on the north slope of Alaska to 43°C in Saudi Arabia. What we experience are the black circles in Figure 37, not the yellow boxes. Whether the yellow boxes are sloping at 0.19°C or 0.14°C, regardless of the cause, doesn't matter much to us, but the 0.14 degree/decade rate is much more likely than the suspicious 0.19 degree/decade rate.

Tamblyn's comment, Happer's reply

Another very odd statement is from Tamblyn's Final Reply:

> "The Professor appears to think CO_2 is the only factor influencing climate; thus, if climate does not respond to CO_2 levels alone, this is evidence against the role of CO_2." (Tamblyn, 2021b, p. 9)

Happer clearly does not think "CO_2 is the only factor influencing climate" as Tamblyn would have known, had he carefully read Happer's Interview:

> "There has been warming, about 1°C, but with half the warming occurring before there was much increase in CO_2. The erratic nature of the warming over the past century suggests that half or more of the warming is not due to more CO_2, but has been caused by other natural phenomena, which have caused warming or cooling over all of geological history." (Happer, 2021a, pp. 38-39)

Later in Tamblyn's Final Reply, we see this:

> "If climate sensitivity [CS] is low, how do we explain such a highly variable climate? A climate system with a significant CS and a significantly changing greenhouse effect explains much of what we see. CO_2 has been called the 'main control knob' for the climate, and for good reason. The Professor's personal view of a low CS, unsupported by any evidence from him, does not match the observations." (Tamblyn, 2021b, p. 14)

We just saw that he criticized Happer for thinking "CO_2 is the only factor," then he claims CO_2 is the "main control knob."[97] So, Happer can't explain a highly variable climate because he believes the climate sensitivity is low? Maybe it is because he thinks other factors are more important than CO_2. As we will see in this book, especially in Chapter 16, Happer provides a lot of evidence that CS is low.

Tamblyn was not as polite and thoughtful as Karoly and, sometimes stained the debate, which was supposed to be a "civil dialogue." Tamblyn's arguments are not well thought out and like those used by the press and politicians. The arguments have little scientific content and generally rely on appeals to authority, name-calling, strawman arguments, "red herrings," or ad

---

[97] The "control knob" phrase is probably from a paper by Andrew Lacis and colleagues, which has the phrase in its title (Lacis, Schmidt, Rind, & Ruedy, 2010).

hominem attacks. But we will incorporate many of his comments in this discussion, so all views, good and bad, are presented and discussed.

As Happer explains in his Interview, from page 23 to 29, current surface temperature estimates are heavily massaged with numerous corrections and adjustments. The adjustments to HadCRUT5 have removed the "hiatus in warming that was clear in the data of 2013." (Happer, 2021a, p. 25). Happer equates this to George Orwell's hero, Winston Smith, in *1984*, who is tasked with rewriting history to conform to the ruling party line. Next, we visit the Pause—or hiatus—in warming from roughly 1998 to 2013.

Chapter 7: The Pause

The famous Pause in warming from 2000 to 2013, is marked on the global UAH lower-atmosphere-average temperature plotted in Figure 39. As mentioned above, and by Happer in his Interview on page 25, it has been "adjusted" out of most surface temperature records such as HadCRUT5.

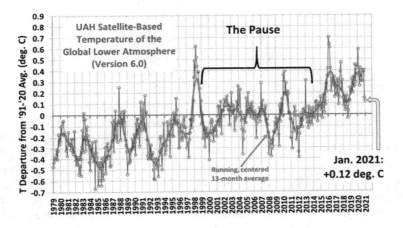

Figure 39. UAH global average temperature of the lower atmosphere through January of 2021. Source: Dr. Roy Spencer (Spencer R. , 2021)

Various lengths have been proposed for the Pause, but it is fifteen years long in the UAH record, even when ignoring the strong 1998 El Niño and the beginning of the 2016 El Niño. Happer places the length at 14 years, from 2000 through 2013. The Pause deeply embarrassed the climate establishment because they claimed that CO_2 was the "climate control knob"[98] and, yet CO_2 concentration was rising quickly, while temperature did nothing. Obviously, there were natural climate influences that could override CO_2-caused warming for fifteen years, so how could CO_2 be the control knob? The IPCC had this to say about the Pause in AR5:

> "Almost all CMIP5 historical simulations do not reproduce the observed recent warming hiatus. There is medium confidence that the GMST trend difference between models and observations during 1998–2012 is to a substantial degree caused by internal variability, with possible contributions from forcing error and some CMIP5 models overestimating the response to increasing GHG and other anthropogenic forcing." (IPCC, 2013, p. 772)

---

[98] (Lacis, Schmidt, Rind, & Ruedy, 2010)

Changing the data to match the theory

Next, Happer brought up a paper by Thomas Karl and colleagues[99] in which they adjusted surface measurements in such a way as to lower earlier temperatures and raise later temperatures. Their new adjustments were just enough to erase the Pause. In the paper, it was referred to as the "hiatus" in warming and their changes to the surface-temperature record to hide it were widely ridiculed as "Karlizing" the data (Watts A. , 2016b).

Background: Global temperature adjustments

But the massive effort to change the data to match the theory was not over. The satellite-temperature records still showed a pause (see Figure 39). Carl Mears and Frank Wentz[100] of Remote Sensing Systems (RSS), a competitor to the UAH satellite-temperature dataset, introduced their own set of corrections that were meant to "erase" the Pause. Essentially what they did was reintroduce some bad, anomalously warm data from the NOAA-14 satellite to the mix used in their previous record. Roy Spencer at UAH commented that NOAA-14 required a large correction because it was drifting into a new orbit. Throughout the same period, data from NOAA-15, operating with a more advanced microwave detector and requiring no correction, was not showing any warming. Spencer then tells us that RSS "decided to force the good data from [NOAA-15] to match the bad data from NOAA-14" (Watts A. , 2016b).

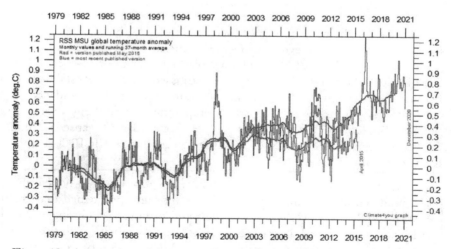

Figure 40. A comparison of the previous RSS satellite temperature record, in red, and the "corrected" record in blue. Source: Climate4you.com.

---

[99] (Karl, et al., 2015)
[100] (Mears & Wentz, 2016)

The Mears and Wentz paper was rejected by the *Journal of Geophysical Research* but eventually accepted by the *Journal of Climate*. The magnitude of their correction can be seen in Figure 40. The original RSS record is shown in red, and the new record is shown in blue. The original record is like the UAH record shown in Figure 39. Figure 40 was added to the Happer Interview as Figure 21 by the editor (Happer, 2021a, p. 27).

What really controls global warming?

Karoly, not surprisingly, has a different take on the Pause. He writes:

"The so-called "hiatus" in global warming was a creation of a small number of commentators who usually considered the period from 1998, a very hot year, to around 2012. Nineteen Ninety-Eight was a very hot year associated with the strongest El Niño event in the last 150 years. The years 2011 and 2012 were colder, being associated with back-to-back La Niña events. It is not surprising that if you select a period starting in a hotter year, like 1998, and finishing in a colder year, like 2012—both associated with natural variability—that there is a reduction in the warming trend in global average temperature due to natural variability.

"Does this mean global warming stopped? No—for two reasons. First, global warming is the addition of heat to the climate system, not just the increase in global average temperature. In fact, about 90 percent of the heat added to the climate system due to the increase in greenhouse gases in the atmosphere over the last 100 years has gone into heating up the oceans and only a small fraction has gone into heating the atmosphere and the land surface. The heat content of the ocean, particularly the deep ocean, has increased more rapidly over the last two decades, even while the global average surface temperature showed a reduced rate of warming due to natural variability from 1998 to 2012." (Karoly, 2021a, pp. 16-17)

He acknowledges that the Pause exists and was caused by natural variability due to ENSO events. Then he defines global warming as the accumulation of heat in the climate system. He uses heat in the colloquial sense, he means thermal energy. Strictly speaking, heat is the transfer of thermal energy from a warm object to a cooler object. We often use heat in the colloquial sense in this book as well.

Background: Earth's surface heat content

The Pause from 2000 through 2014 does not depend upon the 1998 El Niño or the 2010 La Niñas, but other than that, Karoly does a good job with this description. He is correct with his claim that 90% of the heat added to the climate system has gone into heating the oceans, in fact it is probably more than 90%, depending upon what you wish to include in the climate system.

We only have good data on ocean temperatures since 2002–2005, but the trend since then shows warming of 0.06°C/decade. Figure 41 uses yearly data from JAMSTEC to 2,000 meters and University of Hamburg global average temperatures between roughly 2008 and 2020 below 2,000 meters. The data below 2,000 meters is not yearly data, so we are assuming that these deeper temperatures have not varied much between 2002 and 2020. This is likely true since the yearly average temperature at 2,000 meters only varies from 2.44°C to 2.47°C through the same period.

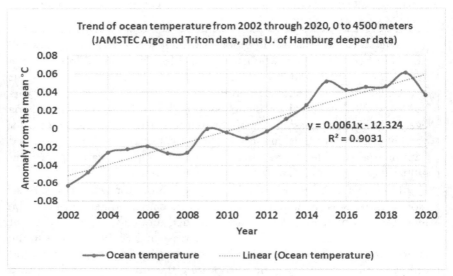

Figure 41. Trend of estimated ocean-temperature anomalies from the surface to 4,500 meters since 2002. Data sources JAMSTEC and University of Hamburg.

We have seen evidence that the surface and lower troposphere are warming between 0.14°C/decade and 0.19°C/decade since 1979. We also have evidence, from HadCRUT5, that the surface is warming at 0.1°C/decade since 1900 and 0.18°C/decade since 1960. Earth's surface is generally defined as the lower two meters of the atmosphere, a small volume of the planet's surface. If we were to define the surface as everything between the ocean floor and the top of the atmosphere, which we take as an altitude of 84 kilometers, we encompass everything influenced by Earth's climate, except the land surface.

Specific heat capacity is defined as the amount of energy that must be added to a substance to raise its temperature one degree. Water has a large specific heat capacity (4.2), which means a lot of energy must be added to increase its temperature. In contrast, air has a low specific heat capacity of 1.0. As a result of this difference, and the fact that the water in the oceans weighs 265 times as much as the entire atmosphere, the oceans contain 99.9% of the total heat stored in the atmosphere and the oceans. As mentioned in the last chapter, this means the heat in the lower two meters of the atmosphere is only 0.00002% of the total.

Determining the magnitude of climate change by estimating the change in surface temperature is a bit silly, we don't weigh a straw to determine how much hay is in the barn, we weigh something larger, like a bale, then count the bales. The surface is not only a tiny part of the whole system, but also where most of the chaotic weather occurs. Why not go to where nearly all the thermal energy is stored, won't the estimate be more reliable? As Figure 41 shows the ocean temperature is rising slowly, only 0.06°C/decade since 2002 and the trend is 90% linear ($R^2 = 0.9031$), making acceleration unlikely. The debate over how fast the surface temperature is rising, or even the temperature of the entire lower atmosphere is insignificant. The oceans are the real control over global warming, as Karoly implies in the quote above.

Background: Solar Anomalies from 1997 to 2000

The Pause started between 1997 and 2000 and there are several indications that the Sun changed during this period. The ACRIM TSI record changes direction, from upward to downward at this time, as discussed in Chapter 4. The solar *aa* index[101] measures certain aspects of the solar wind and tracks sunspots but does not correlate well with them (Feynman & Ruzmaikin, 2014). The index reaches a multidecade low in 1997, this low precedes, and foreshadows, an extreme 100-year low in 2009. Clearly a major secular change in the solar wind occurs around 1997.

Besides possible changes in the Sun, cloud cover fell 4 to 5% from 1985 to 2000 and then began to increase. Cloudiness increased nearly 3% by the end of 2004, contributing to the Pause. Earth's total albedo (reflectivity) hit a low during 1997 and increased afterward.[102] Goode and Pallé found that the albedo decreased 10% from 1984 to 1997 causing a 10 W/m² increase in shortwave solar radiation striking Earth and resulting in probable warming. The nominal albedo is about 30%, meaning that normally 30% of sunlight is reflected by the Earth to space. The roughly 10% drop to 1997 means this would be reduced to 27%.

From 1997 to 2005, the albedo increased 3%, reducing the solar radiation striking Earth by about 3 W/m², resulting in probable cooling. This increase in albedo would increase the albedo from the assumed 27% to about 28%. As Goode and Pallé write in their paper, changes in the albedo are closely related

---

[101] The *aa* index is the antipodal amplitude. It measures differences in the disturbance in Earth's magnetic field at two nearly antipodal stations. One is in the UK and the other is in Australia. The disturbances are caused by solar magnetic field changes brought to Earth by the solar wind.

[102] (Goode & Palle, 2007) measure the variations in solar output and in Earth's albedo to compute the actual radiation that strikes Earth's surface as a function of time from the mid-1980s to 2005. They use land-based measurements and satellite measurements.

to changes in cloud cover. They also write that considering solar changes without considering albedo changes can be misleading. Both affect the insolation received on Earth's surface.

Goode and Pallé remind us that the relationship between cloud cover and warming or cooling is more complicated than simply considering albedo changes. Current theory suggests that lower-level clouds are considered cooling clouds and upper-level clouds are considered warming.

This is discussed in more detail in Chapter 9, but here we will briefly introduce the topic. While increasing the albedo will always reduce the solar energy striking the surface, an increase in upper-level clouds may hold in energy emitted by Earth's surface, especially at night, warming the atmosphere below the clouds. Because the upper-level clouds are very cold they emit very little energy to space. It is the balance of the Earth's surface energy trapped by upper-level clouds, at night, versus the amount they emit or reflect into space that determines if the clouds are net warming, or net cooling. Lower-level clouds nearly always cool the atmosphere below them because reflection dominates at that altitude. Lower-level clouds are almost as warm as the surface and emit nearly as much energy upward.

What is clear is that Earth's recent warming and the Pause in warming are the result of a complex interaction of processes. The changing concentration of atmospheric CO_2 is only one of these and may not be the dominant one. However, regardless of the numerous unknowns in the CO_2 global warming hypothesis, many in the consensus say we should stop or reduce our fossil fuel consumption anyway. This idea is called the "precautionary principle." Keep the complexity and uncertainty just discussed, in mind as you read the next chapter.

Chapter 8: The Precautionary Principle

TheBestSchools.org asks Dr. Happer:

> "Many would argue that whatever the theoretical and empirical uncertainties surrounding the consensus view on global warming may be, we simply cannot afford the luxury of further study. As with Pascal's Wager, the stakes are so high that it is far better to act and discover it was not necessary, than not to act and discover it was.
>
> "…we would like you to speak to this general type of argument: 'urgency overrides normal scientific caution in the face of uncertainty.'" (Happer, 2021a, p. 33)

Happer's reply:

> "Pascal made major contributions to physics and mathematics of importance to climate science, before turning to philosophy and theology. As far as I know, he was the first to introduce probabilistic or statistical ideas to theology. And the theological context is appropriate, since global warming has long since acquired many of the trappings of religion, disguised as science. Promoters of the 'good insurance' argument would have you believe that there is a small but finite risk of catastrophic consequence from more CO_2, irreversible 'tipping points,' and other doomsday scenarios. This is not true. CO_2 levels were thousands of ppm over most of the Phanerozoic Eon."

The Phanerozoic Eon comprises the last 540 million years; it is the period with a good fossil record of multicellular life. Happer then shows a CO_2 reconstruction for the Phanerozoic, like our Figure 16 in Chapter 3, from R. A. Berner and Z. Kothvala's paper[103] (Happer, 2021a, p. 34). Figure 11 shows a reasonable reconstruction of Phanerozoic global average temperatures from the Smithsonian Institution.

CO2 was much higher in the past

Berner and Kothvala's figure shows that, in the past, CO_2 was 10 to 25 times higher than today without Earth reaching a "tipping point" and falling into disaster. Happer then writes in his Major Statement:

---

[103] , "Geocarb III: A Revised Model of Atmospheric CO_2 over Phanerozoic Time" in the *American Journal of Science* (Berner & Kothavala, 2001)

"The important message of Fig. [16, Chapter 3] is that CO_2 concentrations have been much higher than present values over most of the history of life. Even though CO_2 concentrations were measured in thousands of parts per million by volume (ppm) over most of the Phanerozoic, not the few hundred ppm of today, life flourished in the oceans and on the land. Average pH values in the ocean surface were as low as pH = 7.7, a bit lower than the pH = 8.1 today. But this was still far from acidic, pH < 7, because of the enormous natural alkalinity of seawater. The mean global temperature was sometimes higher and sometimes lower than today. But the temperature did not correlate very well with CO_2 levels. For example, there were ice ages in the Ordovician, some 450 million years ago, when the CO_2 levels were several thousand ppm." (Berner & Kothavala, 2001) (Quinton & MacLeod, 2014), (Happer, 2021b, pp. 3-4)

We have no response from Karoly, but Tamblyn responds with:

"Lest we wonder how the Ordovician ice age could have occurred ... GEOCARB III [Figure 16] is a geochemical model, which estimates past CO_2 levels from the chemistry of rocks. Its calculations are run in steps of 10 million years and averaged over 50 million years. It is not sensitive enough to detect shorter-term changes. Direct geological evidence shows that CO_2 levels fell sharply during that period, in 1 to 2 million years or less—too small for GEOCARB III to capture. A higher-resolution geochemical model applied just to this period suggests a decline of CO_2 levels from ~5000 to 3000 ppm. With differences in solar output, 3000 ppm then is equal to 500 ppm today. Climate models applied to late Ordovician conditions predict icehouse conditions at CO_2 levels below about 2240 to 3920 ppm. (Tamblyn, 2021a, pp. 36-37)

We discussed the Ordovician ice age in Chapter 3, but even if 3,000 ppm in the Ordovician ice age is equal to 500 ppm today, we are currently at 400 ppm. Further, the CO_2 concentration during the last glacial maximum was only 190 ppm. Tamblyn's point about resolution is valid but does not prove his case and his numbers are inconsistent with his thesis. He lists many valid influences on climate but fails to make the case that CO_2 is the dominant influence. Further, as noted in Chapter 3, the IPCC claims that CO_2 concentrations above 300 ppm preclude icehouse conditions on Earth (IPCC, 2013, p. 435).

Background: CO_2 levels and impact

Tamblyn tries to suggest the Ordovician ice age was a brief episode, but it lasted at least 20 million years and probably caused a major mass extinction (Sheehan, 2001). It was not a small event. The decrease in CO_2 was probably related to the cooler temperatures at the time, which caused more atmospheric CO_2 to go into solution in the ocean. Several reasons for the Ordovician ice age have been suggested, including a collision with an asteroid (Schmitz, et al., 2019) and the solar system moving in and out of the arms of the Milky Way.[104] CO_2 concentration was too high at the time to have been a factor. Tamblyn's comments do not strengthen his case.

Today's CO_2 concentration is low for many plants. Joy Ward and colleagues studied fossilized juniper-tree fragments in the La Brea tar pits and found that they suffered from carbon starvation in the last glacial period (Ward, et al., 2005). At the time, between 7.7 and 55 kyr, CO_2 concentration was between 180 and 220 ppm. Compare that with 400 ppm today.

David Tissue, J. K. Dippery and colleagues at Duke University found that, at least for some plants, the CO_2 concentration required for survival of the species is about 15 Pa (partial pressure) CO_2, which is equivalent to 152 ppm in the atmosphere.[105]

Plants can be divided into two groups depending upon the way they breathe in CO_2. The two plant groups are called C3 and C4. The highly efficient users of CO_2, C4 plants, first appeared in the evolutionary record around 30 Ma, in the Oligocene Period, just as the current Antarctic ice cap was forming, ushering Earth into the current ice age. The C4 major-dispersal-event occurred between five and three Ma as Earth entered the deepest and coldest period of the current ice age (Sage, 2003). Rowan Sage believes that arid conditions, associated with ice ages, and lower CO_2 concentrations contributed to their evolution.

The newer C4 plants protect their CO_2 processing molecule, called RuBisCO, with a sheath of 4-carbon molecules to keep out O_2 molecules. When CO_2 levels are very low, RuBisCO will attach to oxygen molecules, which causes them to produce poisonous hydrogen peroxide rather than nutritious carbohydrates.

The 4-carbon sheath gives the C4 plants, like sugar cane and corn, their name. All plants grow better with more CO_2, but C3 plants benefit the most because they lack the protective C4 sheath (Kirkham, 2011). Crop yield is a function of many variables, but generally additional CO_2 encourages more plant growth.

This is part of the reason yields of C3 plants, like wheat and rice, are rising so fast as CO_2 concentrations increase. From 1961 to 2018 the world yield of both crops rose about 300% according to OurWorldinData.org. Roughly 85%

---

[104] (Shaviv & Veizer, 2003) (Shaviv, 2003b)
[105] (Tissue, Griffin, Thomas, & Strain, 1994) (Dippery, Tissue, & Thomas, 1995)

of all plants use the C3 photosynthesis process. Besides wheat and rice, most other cereals, cotton, potatoes, and soybeans are C3 plants. Most trees and grasses are also C3 plants. C4 plants include corn and sugarcane, and while corn yield has also increased 300% from 1961 to 2018, sugar cane yield has only risen 150% in the same amount of time.

Dippery conducted tests of the additional growth expected from both groups of plants under elevated CO_2 concentrations. At 35 days, his C3 plants (*Abutilon theophrasti*), grown at 150 ppm, only had 8% of the growth seen at 350 ppm. Growth was 22% higher at 700 ppm than at 350 ppm. More significantly, root mass was 54% greater in C3 plants grown at 700 ppm, 300 ppm more than today. Dippery and colleagues did not see any additional growth with additional CO_2 in their C4 plants, but the C4 plants had 16% less growth at 150 ppm. As we saw with corn and sugar cane, all C4 plants are not the same, how CO_2 concentration affects each plant is different.

Other studies show that even C4 plants use less water per pound of growth under higher CO_2 concentrations since they have smaller and fewer stomata (leaf breathing holes) under higher CO_2 conditions (Taub, 2010). CO_2 does not change the plant DNA to accomplish this, the plants constantly adjust the size and number of stomata on new leaves, depending upon the environment. All plants try and optimize their stomata to both minimize water loss and maximize photosynthesis efficiency.

C4 plants do seem to have greater control on their water use efficiency than C3 plants and appear to adjust their water usage based partially on water availability.[106] Besides requiring less water, most plants require less nitrogen when CO_2 levels are elevated.[107]

The net result of all these efficiencies gained from higher CO_2 levels is a greening planet. Randall Donohue and colleagues have estimated that the 14% increase in CO_2 from 1982 to 2010 has led to a 5 to 10% increase in green foliage cover in warm, arid environments. Analysis of satellite photos reveals that actual plant cover increased 11% over the same period in these environments.[108]

The Debater's discussion of the risks

Happer's argument against the precautionary principle is that more atmospheric CO_2 has always been better for life on Earth and will likely be good for us in the future. There is no need to do anything, preemptively or otherwise, if the evidence says there is no problem. His secondary argument is that there is no way for humanity to prosper without using fossil fuels. Quite aside from the high cost of solar and wind energy, they are also unreliable.

---

[106] (Polley, Johnson, Mayeux, Brown, & White, 1996)
[107] (Taub & Wang, 2008)
[108] (Donohue, Roderick, McVicar, & Farquhar, 2013)

Happer fears the least advantaged among us will suffer the greatest devastation from the precautionary principle.

Karoly does not directly address the precautionary principle, but Tamblyn does. His argument is that the climate system has high inertia, it resists change. In his view we are "locking in" future changes to sea level and other parts of the climate today that will commit "our descendants to an uncertain, and possibly very harsh, future." (Tamblyn, 2021b, p. 22).

Karoly emphasizes that "global warming is a very long-term problem." He predicts that it will affect humanity for more than 1,000 years. Karoly wants us to consider future generations when we consider the costs and benefits of global warming and CO_2 emissions (Karoly, 2021a, p. 32).

In Tamblyn's view, he has demonstrated that continuing to release GHGs is not safe. He also believes that Happer has not demonstrated that continuing to release GHGs is safe. Tamblyn acknowledges that halting the use of fossil fuels entails considerable risk. The world will suffer significant economic harm by switching to more expensive and unreliable green-energy sources. He appears to acknowledge that this is the case now, but holds out hope that, in the future, the costs of renewable power will be less.

Oddly, he seems to be saying that Happer must prove that CO_2 and burning fossil fuels are not controlling global warming. This seems an unwarranted shift of the burden of proof since nature has always controlled the climate in the past and human influence on the climate has never been observed or measured. Tamblyn also seems to think that even though he has not proven CO_2 controls the climate, we should get rid of fossil fuels anyway, and hope that cheaper alternatives appear in the future. None of this seems "safe" or sensible.

Future costs in perspective

Happer's discussion of global warming, CO_2 emissions, and their impact tends to emphasize what we can measure today. Happer says the current additional CO_2 and warming are clearly beneficial right now. That said, why should this change in the future? The world is getting greener as plants move into areas where they could not grow decades ago. Farmland has higher yields in warmer weather with more available CO_2, and the warmer climate has opened more land to farming. Dr. Craig Idso has calculated a direct monetary benefit to the world of US$3.2 trillion over the 50-year period from 1961 to 2011. He expects this CO_2 windfall to grow to $13 trillion by 2050 (Idso C. , 2013).

Karoly and Tamblyn, as well as the IPCC, tend to discuss projected impacts on the world from additional CO_2 and global warming. They also claim we are seeing climate impacts today, as we will address in later chapters, but the evidence is controversial and thin. Sometimes they discuss the far future—as much as 1,000 years. They do these projections without considering the new technologies that will be developed and the likely growth in the world economy. Imagine the world of 100 years ago, compared to today?

Background: Economic impact of climate change

Global gross domestic product (GDP) was about US $88 trillion in 2019 and it is growing about three percent per year. This means that in 2100, when we might see some adverse effects of global warming, according to the IPCC, the global GDP is likely to be about US $909 trillion, an increase of ten times. They will have ten times the money we have now, in today's dollars, to deal with whatever comes up, if anything.

Bjørn Lomborg, a climate-change economist has estimated that the cost of climate change today is close to zero, specifically he writes: "global warming causes about as much damage as benefits." This estimate is supported by Richard Tol and many other economists (Lomborg B. , 2016). Climate related deaths have fallen 99% in the last century and the cost of weather-related disasters as a percentage of GDP has also fallen dramatically[109].

Tol explains that, since we cannot measure any climate change impact today, we must assume some warming scenario and its associated cost (Tol R., 2009). He gathered and analyzed estimated costs from several projections and found, collectively, they suggest no significant negative impacts are likely until the global average-surface temperature reaches two degrees warmer than in 2009.

Lomborg also looked at the potential costs of global warming versus the cost of radically reducing CO_2 emissions in a 2020 paper in *Technological Forecasting & Social Change* (Lomborg B. , 2020). In a detailed analysis of economic projections by Nobel Prize winning Yale Professor William Nordhaus, the IPCC, and others, Lomborg also finds the change in global GDP today due to climate change is imperceptible. He projects a slight change of -0.28% in GDP at 1.5°C of warming, and a -2.9% impact in 2100 due to the 4°C of warming expected if we do nothing about climate change.

Lomborg points out that the IPCC projects an increase in human welfare of 450% between now and the end of the 21st century. Projected climate damages, if we do nothing, would reduce this gain in welfare to 434%. Thus, the IPCC projects climate change to reduce the gain in global wealth by a little over 3% in 2100.

Lomborg goes on to point out that, while the cost of doing nothing is small, the costs of proposed climate policies are very large. The Paris agreement, if fully implemented, will cost over $800 billion per year in 2030 and will reduce emissions by just 1% of what is needed to limit the average global temperature rise to 1.5°C. Each dollar spent on Paris will result in only 11¢ of benefits. Not a very good deal.

---

[109] (Clintel.org, 2021), (Pielke Jr., 2017), and (Lomborg B. , 2020)

When and if climate change causes any serious problems for humankind depends upon how much we believe CO_2 affects global warming and which CO_2 emissions scenario we use. The more extreme scenarios, such as the commonly used RCP8.5 IPCC "business as usual scenario," have been shown to be highly unlikely,[110] so we will examine CMIP6 projections using the moderate emissions scenario SSP245 plotted in Figure 42.[111]

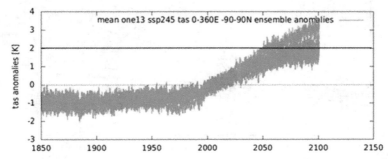

Figure 42. CMIP6 SSP245 projections of temperatures to 2100. The anomalies are relative to the 1991 to 2020 average and reach two degrees above 2009 around 2100. 13 model runs are plotted. Source KNMI Climate Explorer.

All IPCC AR5 and AR6 CO_2 emissions scenarios fall in the gray area of Figure 43. The scary scenarios used by the news media and politicians are the implausible business as usual scenarios labeled RCP8.5 (AR5) and SSP5-8.5 (AR6) shown by the upper black and orange lines, respectively.

More reliable emissions projections by ExxonMobil, BP, the U.S. Energy Information Agency (EIA) and the International Energy Information Agency (IEA) cluster around and below the AR6 SSP2-4.5 scenario, which is the dark red line near the bottom of the gray area in Figure 43. The independent assessments are circled in blue. Clearly most of the IPCC CO_2 emissions scenarios are much higher than other sources.

In summary, the precautionary scenario does not apply here, according to Happer, because our geological past tells us we've had more than ten times the CO_2 concentrations foreseen for the future many times throughout Earth's history and nothing catastrophic happened. Further, the higher CO_2 concentrations we see today are beneficial or neutral so far. No observations suggest any problems now or in the years ahead.

---

[110] (Ritchie & Dowlatabadi, 2017) and (Burgess, Ritchie, Shapland, & Pielke Jr., 2020)

[111] Figure 42 plots the model results of the KNMI Climate Explorer thirteen-member model ensemble, which includes models from Canada (Canadian Centre for Climate Modeling), the U.S. (NCAR, NOAA, U of Arizona), Germany (Alfred Wegener Institute), China (Beijing Climate Center, the Chinese Academy of Meteorology, and the Chinese Academy of Sciences), Japan (JAMSTEC), and the U.K. (Hadley Centre). Some of the participants submitted multiple runs of their models.

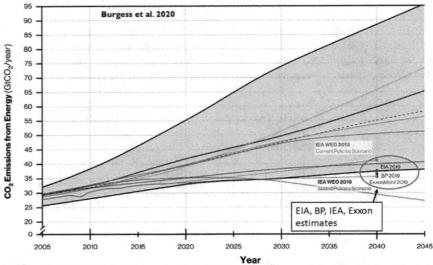

Figure 43. The IPCC AR5 and AR6 CO2 emissions scenarios. All model results fall within the gray area. The upper orange and black lines (within the gray) are the AR6 SSP5-8.5 and the AR5 RCP8.5 emissions scenarios, respectively. The EIA, IEA, BP, and ExxonMobil projections are within the blue circle. These 2040 projections fall around the red SSP2-4.5 line, used in Figure 42. Source: (Burgess, Ritchie, Shapland, & Jr., 2020).

Tamblyn and Karoly rely solely on IPCC projections to support their assertions that we must do something now to prevent possible danger far in the future. Yet, basic economics shows this to be wrong. Our economy is ten times smaller today than it will be in 2100, and we would be spending far more today, in terms of economic harm, than the spending might save our descendants in 100 years. Further, 100 years from now we will be more technologically advanced and better prepared to deal with any problems that might arise, if any.

The IPCC global average temperature and CO_2 emissions predictions are greatly inflated. Current observations suggest CO_2 emissions—if they affect the climate at all—are trending near the lowest of the IPCC scenarios seen in Figure 43. We cannot be sure that CO_2 emissions have any impact on climate, and we are benefiting from increasing CO_2 in our atmosphere today. The precautionary principle does not apply.

Chapter 9: Do Humans and CO$_2$ drive Climate Change?

Karoly writes the following:

> "The observed large-scale increase in surface temperature across the globe since the mid-twentieth century is primarily due to human activity, the increase of greenhouse gases in the atmosphere, and other human impacts on the climate system." Karoly Interview (Karoly, 2021a, p. 40)

Karoly explains that the recent increase in atmospheric CO$_2$ is associated with a decline in the ratio of the isotopes ^{13}C to ^{12}C (Keeling, et al., 2017), which is expected if some of the CO$_2$ is from burning fossil fuels since plants prefer ^{12}C and most fossil fuels are generated from decayed plants. In addition, there is a slight decrease of atmospheric oxygen as one would expect from burning fossil fuels.

Karoly adds:

> "The increase in atmospheric carbon dioxide over the last 40 years agrees very well with the increase expected from emissions associated with burning fossil fuels, land clearing, and industrial activity, less the additional uptake of carbon dioxide into the oceans and the land ecosystems due to the higher concentrations." Karoly Statement (Karoly, 2021b, p. 11)

Happer agrees that the observed increase in atmospheric CO$_2$ concentration is due to human activity. He just disagrees it is necessarily related to dangerous global warming.

Happer writes:

> "Most of the [recent] warming has probably been due to natural causes. But much of the increase in CO$_2$, from around 280 ppm in the year 1800 to about 400 ppm in 2015, is probably anthropogenic, although the warming oceans and land have also released some CO$_2$. The warming of urban areas has correlated well with increasing CO$_2$. This is the well-known urban heat-island effect of expanding cities. But it is not increasing CO$_2$ that causes urban warming; rather, it is the replacement of green fields and forests, with their transpirational cooling, by roads and buildings which do not transpire water vapor. Of course, CO$_2$ levels also increased, along with the urban warming, but the additional CO$_2$ did not cause the warming associated with urbanization. ... Over the non-urban areas of the earth, the correlation between CO$_2$ levels and temperature has been poor." Happer Interview (Happer, 2021a, p. 18)

Karoly also points out that during the last 800,000 years of the Pleistocene Epoch, Antarctic ice cores suggest that CO_2 levels have never been above 300 ppm. Thus, the current level of 400 ppm is very unlikely to have a natural cause, such as volcanic eruptions or CO_2 out-gassing from the warmer oceans.

Most of the additional atmospheric CO_2 is from humans

There is agreement between Happer and Karoly that a portion, perhaps most, of the increase in CO_2 concentration in the atmosphere is due to human activity. There is also agreement that about 40% of the CO_2 humans add to the atmosphere is taken up by land ecosystems and the oceans.[112] This is easily seen in the Hawaii CO_2 record begun by Charles Keeling in 1956 (Scripps Institute of Oceanography, 2021) where the seasonal oscillations in CO_2 have increased in magnitude as more plants are added to the Northern Hemisphere, partly because of CO_2 fertilization.

As Matt Ridley has written, during Northern Hemisphere summers, the Earth breathes in carbon dioxide as the seasonal increase in plant life absorbs the gas while growing. In the winter, Earth breathes out CO_2 as the leaves of summer fall to the ground and rot (Ridley, 2013b). In 1985, Keeling noticed this yearly cycle was increasing in amplitude from 13.3 ppm in 1969 to 14.5 ppm in 1981. He interpreted this increase as meaning the total plant life in the Northern Hemisphere increased during that period.[113]

As mentioned in the last chapter, numerous papers have been published over the past ten years that use satellite photography to show that Earth is greening. An important 2016 paper[114] found that the growing season increased in 25 to 50% of the global vegetated area from 1982 to 2009. Less than four percent of the globe has shown a loss of leaf area (browning). They calculated that 70% of the observed global greening is due to additional carbon dioxide in the atmosphere. NASA has estimated that the increase in leaf-covered area is equivalent to two times the area of the continental United States (Reiny, 2016).

As Happer notes, humans paving over urban areas and roads reduces the local plant uptake of CO_2. However, the additional CO_2 increases the growing season and the amount of plant life in other areas. A separate study, also published in *Nature*, in 2019, showed that urban areas have longer growing seasons than nearby rural areas.[115] They showed that the increased urban growing season was sensitive to both additional CO_2 and higher urban

---

[112] (Keeling, Whorf, Wong, & Bellagay, 1985)

[113] (Keeling, Whorf, Wong, & Bellagay, 1985)

[114] *Nature Climate Change* by Zaichun Zhu and colleagues (Zhu, Piao, & Myneni, 2016)

[115] (Wang, Ju, Penuelas, Cescatti, & Zhou, 2019)

temperatures. We will discuss the impact of additional CO_2 more in the next chapter.

While Happer and Karoly agree that CO_2 is increasing, due to human activities, they disagree about the impact of the increase. Karoly thinks the increase will lead to dangerous global warming, but Happer thinks it will lead to still more greening of the planet, more arable land, and more abundant food.

As Happer notes, the warming in urban areas and in climate models correlates better with CO_2 than the warming in rural areas. Willie Soon and colleagues examine this topic in some detail.[116] Next, we ask how warming is attributed to humans.

Attributing warming to humans

TheBestSchools.org ask David Karoly in his interview: "If increased atmospheric CO_2 is not responsible [for global warming], what is?" TheBestSchools notes that skeptics point out several factors that need to be considered before we can reach this conclusion:

1. We are, after all, still recovering from the last ice age (in the true sense of the term), which lasted for about 100,000 years and only ended about 12,000 years ago; therefore, why isn't modest warming simply what we should expect (the null hypothesis) ...
2. Over the geological record taken as a whole, it appears that warming trends regularly precede rising CO_2 levels, not the other way around.
3. Some studies show a strong correlation between solar activity cycles and Earth surface temperatures. (Karoly, 2021a, p. 21)

Karoly is confident that CO_2 and other "greenhouse gases from human activity" are the major cause of observed warming (Karoly, 2021a, p. 23). He points to the AR5 IPCC report. Specifically, he directs us to Figure 10.5 on page 884 in Chapter 10, which is our Figure 44.

As discussed in Chapter 5, the IPCC computes the human contribution to climate change with models, seven of the fifteen models are listed on the left of Figure 45. Since the human influence on climate has never been observed or measured, the "climate forcings" plotted in Figures 44 and 45 depend solely upon assumptions built into the models. Figure 44 is a summary of the data and analysis illustrated in Figure 45.

---

[116] in a 2018 paper in *Earth-Science Reviews* (Soon, et al., 2018) and in (Connolly et al., 2021)

Contributions to observed surface temperature change over the period 1951–2010

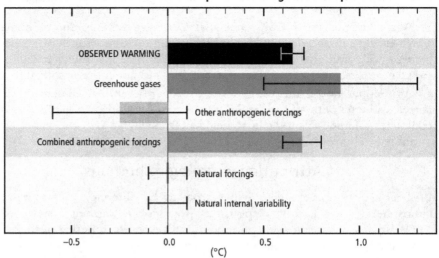

Figure 44. The IPCC modeled ranges for attributable warming trends from 1951 to 2010. Observed warming of about 0.66°C is shown in black. IPCC AR5 (IPCC core writing team, 2014, p. 6).

Background: Analyzing the IPCC models

Figure 45 is a portion of IPCC AR5 Figure 10.4 (IPCC, 2013, p. 882). It displays the modeled greenhouse gas (GHG) caused warming, in green. The left panel, panel (a), is scaled in degrees C, and shows the range of modeled changes in temperature, from 1951 to 2010. The bars shown are based on a regression analysis of numerous computer-runs of each model. The individual computer runs systematically change the conditions, for example some runs have little change in CO_2 and some have large changes in CO_2. The central line in panel (a) is zero change for the analyzed climate forcing.

The forcing color schemes in Figures 44 and 45 are the same, so green represents GHG caused changes, yellow is other-anthropogenic, and blue is natural variability. The green bars vary from negative values (GISS-E2-H) to over 1.5°C (GISS-E2-R and H, CSIRO-Mk3-6-0). The other anthropogenic or OA estimates are just as variable. The most likely results from several statistical analyses of model results are shown as black squares, diamonds, and triangles in panel (b).

Panel (b) shows the same data, but as a scaling factor. While the proportion of warming caused by each factor is model based, the total warming is constrained by the HadCRUT4 record. Thus, the coefficients used to compute AGW, and natural warming are computed from model results, summed, and then forced to match HadCRUT4 (Gillett et al., 2013). Nathan Gillett and colleagues observed that some of the 15 models they studied produce very anomalous negative scaling factors, as shown in Figure 45(b). Their assessment of the model results is shown as triangles in Figure 45(b). The negative scaling factors are a result of negative warming coefficients. This means the component is cooling as its forcing increases. They single out model GISS-E2-H and note the model is not "well constrained."

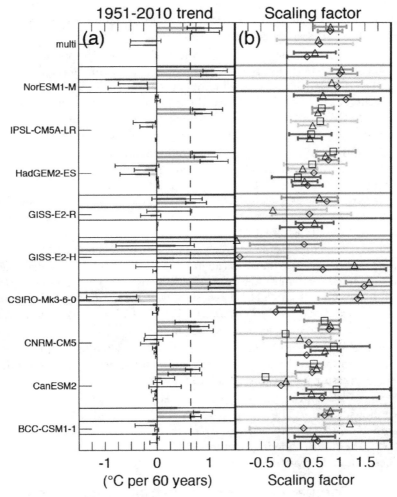

Figure 45. A portion of AR5 Figure 10.4. In panel (a), the green bars are the estimated warming due to GHGs for the model listed to the left. The yellow bars are "other anthropogenic" in Figure 44. The small blue bars are natural forcing. Panel (b) is the scaling factor that must be applied to the model results to match the HadCRUT global temperature dataset. The vertical dashed line in (a) is 0.66°C, the observed warming, and in (b) it is a scaling factor of 1. Source: AR5 (IPCC, 2013, p. 882).

Gillett, et al. comment that: "The assumption is usually made that a model's TCR is proportional to its GHG-induced warming trend over the historical period." GISS-E2-H appears to violate this assumption, which concerns Gillett, and colleagues. They observe that since the desired result is a multi-model estimate of climate sensitivity, the model's violation of the *assumption* should be investigated. Thus, they admit that the connection between GHGs and warming is *explicitly* assumed.

The scaling factor is the amount each model component must be multiplied by, while incorporating its model coefficients, to achieve the best match to the HadCRUT4 global temperature record. The central dashed line is a scaling factor of one, such that the sum of the model components equals the observed warming over the period. Most of the models clearly overestimate warming since most of the scaling factors are below one.

The top set of results, labeled "multi," are multi-model-mean values. The multi-model mean overestimates warming relative to HadCRUT4 for all forcings. Yet, the IPCC believed it was valid to create it from these diverse values and use the average to "compute" the amount of human-caused global warming. The IPCC summary assessment of the values from all 15 models is what is shown in Figure 44.

Some of the models did not detect any GHG forcing (IPCC, 2013, p. 882), but most did. An IPCC statistical analysis of the models suggests that: "Overall there is some evidence that some CMIP5 models have a higher transient response to GHGs and a larger response to other anthropogenic forcings ... than the real world (*medium confidence*)" (IPCC, 2013, p. 884). This conclusion agrees with Happer's assessment of the models.

Figure 45 shows the same thing. Most of the scaling factors, for the seven models shown, are below unity, and some are negative. These models are clearly overestimating warming due to GHGs.

Figure 45 shows an exceptionally large variation in the computed greenhouse effect. The range of computed (or modeled) greenhouse gas warming from 1951 to 2010 is larger than the total observed warming of 0.66°C, the dashed line in panel (a). This does not give us confidence in the values plotted in Figure 44 or in the models.

Comparing the pre-industrial period to today

David Karoly does not believe that we are "recovering from the last Ice Age" (Karoly, 2021a, p. 23). He thinks the long-term trend is a decrease in temperatures over the past 12,000 years and that human CO_2 emissions arrested that trend and reversed it.

Background: Warming trends

The long-term decline in Northern Hemisphere temperatures began in the Mid-Holocene Transition, about 4000BC, as shown in Figure 8, in Chapter 2.

Global and especially Northern Hemisphere temperatures increased or were flat from 10,000 BC to 4000 BC. Karoly, and most experts, think the decline, called the Neoglacial period, is due to Earth's falling orbital obliquity (May, 2016k). As obliquity falls, the tropics receive more insolation at the expense of the higher latitudes. He believes the MBH99 Hockey Stick and thinks that recent warming is unusual, at least over the past 2,000 years, if not longer. Karoly points out that normally warming is different in the Northern and Southern Hemispheres, but in the twentieth century, both warmed.

As discussed in Chapter 6, the Northern Hemisphere is warming faster than the Southern. We recall that the satellite data shows a difference of 0.05°C/decade, and the HadCRUT5 surface data shows a difference of 0.17°C/decade. The difference between the satellite and surface hemispheric trends is large, and even though one is at the surface and the other is higher in the troposphere, it is likely that one of the datasets is in error. The HadCRUT5 Northern Hemisphere warming rate is 2.5 times greater than the Southern Hemisphere's. This difference is possible but suggests that the cause of warming in the North is not the same as the cause in the South, as the IPCC claims. If CO_2 were the cause, and it is evenly distributed globally, how could the warming rates be so different? The warming pattern suggests an orbital or other natural cause—or an error.

AR6 recognizes this spatial warming pattern problem and discusses it in Chapter 7 (IPCC, 2021, pp. 7-88 to 7-91). They acknowledge their models show a very different pattern of warming, especially in the Pacific and Southern Oceans, than observed. However, rather than concluding the models are wrong, or that CO_2 is not driving warming, they conclude that warming is changing the CO_2 feedbacks in some unknown manner.

The National Weather Service Oceanic Niño Index (ONI) dataset was used to count ENSO events since 1980. Based on an ONI cutoff of 1.0°C (El Niño) and -1.0°C (La Niña), there were 57 El Niño months and only 46 La Niña months throughout the period (National Weather Service, 2021). More importantly, the strongest events, of either type, since 1980, were the El Niños in 1982/1983, 1997/1998, and 2015/2016. Perhaps the extra Northern Hemisphere warming can be attributed to the strong El Niños?

The Northern Hemisphere and tropical temperatures are more strongly affected by these events than the Southern Hemisphere. In periods without these strong events, such as from 2003 to 2006 or 2012 to 2014, the hemispheric and tropical temperature anomalies tend to cluster together (see Figure 9 in Chapter 2).

The ENSO oscillations are affected, in part, by the solar cycle as illustrated in Figure 46 originally published by Javier Vinós in a Climate Etc. web post in 2019. His idea is based upon work by Robert Leamon (NASA) and Scott McIntosh (High Altitude Observatory, Boulder, Co.), who were the first to suggest terminations of the solar cycle correlated well with some ENSO events.[117]

Both Vinós, and Leamon and McIntosh successfully predicted the 2020 La Niña years before it happened. Their work suggests that the basic ~11-year solar cycle influences tropical and Northern Hemisphere weather through the ENSO cycle. Not all La Niñas coincide with the ends of solar cycles, but they do tend to occur in a predictable pattern around those times: La Niña-El Niño-La Niña. The Sun and the state of the ENSO cycle could account for some of the observed extra Northern Hemisphere warming.

Karoly believes that solar output has declined for the past 40 years and can be "confidently" excluded as a contributor to recent warming (Karoly, 2021a, p. 24). He also believes the model results are accurate and there is sufficient evidence to reject any significant natural warming, at least from 1951 to 2010.

Figure 46 suggests otherwise, so his view is controversial. Recall from Chapter 4, Figure 20, there are well-supported alternative views that suggest solar output increased until about 2000 and may have significantly contributed to global warming from 1951 to 2000. The debate over whether the Sun's output increased or decreased from the 1970s to the early 2000s is complex, but real.

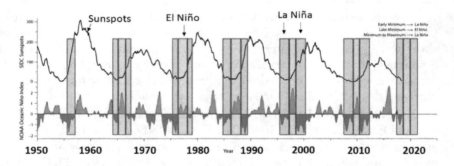

Figure 46. Recent ENSO events are shown in red (El Niños) and blue (La Niñas) along with the SILSO sunspot record. Source: Climate Etc. (Vinós, 2019)

Climate models are the main reason Karoly thinks human activities are driving climate change, but he also appeals to the consensus.

"An overwhelming consensus of climate scientists agree that human-caused climate change is happening, and that global warming will continue throughout the current century, with many adverse impacts on human and natural systems." (Karoly, 2021b, p. 3)

---

[117] (Leamon & McIntosh, 2017) and (Leamon, McIntosh, & Marsh, 2020)

Happer's detailed response contains his view of consensus science.

"Truth has never been determined by 'an overwhelming consensus,' and in fact, consensuses have often been completely wrong." (Happer, 2021c, p. 2)

We will have much more on the debater's views of the consensus in Chapter 11.

In his statement, Karoly dives straight into a discussion of how we know much of the increase in CO_2 concentration in the atmosphere is from humans. On that, Happer agrees. However, Happer and Karoly disagree on the IPCC assertion that CO_2 and other GHGs have caused *most* of the recent warming. The disagreement is over *how much* humans are affecting climate change, not *if* we are.

As discussed in Chapter 5 and earlier in this chapter, Karoly and the IPCC use climate models to compute the greenhouse-gas contribution to global warming.

"Climate model simulations have been used to assess the relative importance of different forcing factors on the climate system and how well they explain the observed global warming. The simulations are driven by natural forcing factors, such as changes in solar radiation and volcanic aerosols, as well as human-caused changes in greenhouse gases and human activity-related climate forcing factors, including industrial aerosols and land use change." (Karoly, 2021b, p. 12)

Karoly refers to the whisker plot in Figure 44 to show the impact of human and natural climate forcings as computed by the climate models from 1951 to 2010. However, as should be clear by now, this plot only reflects the data and parameters fed into the models. Worse, the IPCC model output used to generate Figure 44, is all over the place, as shown in Figure 45.

Proof of human-caused global warming

Karoly also provides the following illustration, Figure 47, published by Skepticalscience.com illustrating how he knows humans are causing global warming (Cook J. , 2011). Figure 47 is Karoly's Figure 9, on page 16 of his Major Statement.

Many of the observations listed in Figure 47, are related to human CO_2 emissions, which both Happer and Karoly agree are the main source of the rise in atmospheric CO_2. The principal disagreement between them is how much warming is caused by the additional CO_2.

We will address the points listed in Figure 47, starting on the upper left and work our way around clockwise.

Background: Shrinking Upper Atmosphere

We assume the upper atmosphere referred to is the ionosphere, which extends from about 80 km to around 1,000 km above the surface. It is mainly influenced by the Sun, and only to a minor extent by other forces. Skepticalscience refers to an article by Jan Laštovička in the *Journal of Atmospheric and Solar-Terrestrial Physics* (Laštovička, 2006) stating that the minor forces include tides, storms, gravity waves, aircraft, and potentially, nuclear bombs. There is no mention of greenhouse gases, and the article does not mention a shrinking ionosphere.

Figure 47. Reasons that the IPCC and Karoly "know" humans are causing global warming. Source SkepticalScience.com.

Background: less heat is escaping to space

The next item is "less heat is escaping to space." This is not controversial, we know more heat is being retained by Earth's surface, because the surface is warming. This fact does not tell us why, only that it is. It does not tell us that humans are causing the warming.

Background: Rising tropopause

Likewise, rising temperatures cause the tropopause to rise. This is not controversial or disputed. This is the reason the tropopause is higher over the tropics than over the poles. Again, this does not tell us why the surface is warming, only that it is.

Background: More fossil fuel carbon in the air

More fossil-fuel CO_2 is in the air. The carbon-isotope ratio discussed at the beginning of the chapter tells us this. It is not disputed and does tell us that human emissions are going into the atmosphere and increasing the CO_2 concentration. It does not establish that CO_2 is causing global warming, but it does suggest it might be a contributor. It does not mean the warming or additional CO_2 are bad things.

Background: More heat returning to Earth

More heat returning to Earth is the next obvious fact mentioned by Karoly and skepticalscience.com. They refer to a paper by W. F. J. Evans and E. Puckrin[118] where Evans and Puckrin use a model, combined with measurements of downwelling radiation, to estimate the additional GHG radiation since pre-industrial times. The measurements were made under clear skies at Trent University in Peterborough, Ontario. They were compared to a model of an assumed pre-industrial atmosphere and assumed that the water vapor concentration has not changed. They estimated the radiation has increased 3.52 W/m^2, which is 40% larger than the IPCC models estimate.

An increase is not unexpected, since the world is warming, nor is this somewhat high estimate large enough to be alarming. According to the Stefan-Boltzmann equation, introduced ahead in Chapter 16, an additional 3.5 W/m^2 should be expected to cause about 0.6°C(1.1°F) of warming, which is so small it might be missed. Some evidence exists suggesting that part of this energy imbalance is caused by additional GHGs in the atmosphere. We should emphasize that water vapor is the most powerful greenhouse gas and their assumption that it has not changed in the past 150 to 250 years is a very large assumption.

Background: Pattern of ocean warming

The next item is "Pattern of ocean warming." They refer to an article in *Science* by Tim Barnett and colleagues (Barnett, et al., 2005) by way of explanation. In the paper, the authors run both natural-forcings and all-forcings climate models in attempts to duplicate the warming patterns observed in the shallow ocean. The models and input used are not explained in the paper or in the supplementary materials, but we can assume they are like

---

[118] 18th Conference on Climate Variability and Change (Evans & Puckrin, 2006)

those used in the IPCC AR4 report (IPCC, 2007b), which were quiet-Sun models (recall Figure 21 from Chapter 4). They assumed that the Sun and ocean cycles are a net-zero forcing over the studied time frame and all the observed additional warming is due to human-emitted GHGs. The logic is circular, if you assume no natural warming, all warming is anthropogenic.

AR6 mentions that the observed patterns of warming in the Pacific and Southern Oceans do not match the models very well (IPCC, 2021, p. 7-88). In AR6 Figure 7.14, page 7-193, they show the pattern mismatch, which covers most of the eastern Pacific Ocean and a lot of the Southern Ocean. In these areas, the models overestimate SST warming, from 1870 to 2019, by up to two degrees per century. The models predict decreasing cloud cover over the same area, but observations show it has increased. Contrary to Figure 47, the pattern of warming suggests the models are not working and factors other than CO_2 are important. We have not reproduced Figure 7.14 here due to the watermark over the figure in the so-called final release of AR6. They should not have released a rough draft; it is annoying and unprofessional.

Background: More fossil fuel carbon in corals, trees, and the ocean

The next three items tell us there is more fossil-fuel carbon in the ocean, in corals, and in trees. Again, as with the atmosphere, this is deduced from the decreased ratio of ^{13}C to ^{12}C seen in each. Fossil fuels are richer in ^{12}C than the atmosphere surrounding them. It is a statement of the obvious, the more fossil-fuel CO_2 in the air, the more winds up in the ocean, plants, and animals. Plants prefer ^{12}C, so they presumably are happy to get more of it. These measurements tell us nothing about the cause of global warming.

Background: Nights/Winters warming faster than Days/Summers

Karoly believes that since winter is warming more than summer and nights more than days, this shows "that changes in solar radiation cannot be a major contributor to [recent] observed warming trends" (Karoly, 2021b, pp. 15-16). If he limits this conclusion to the direct effects of solar radiation on the surface, this is very likely true. But the Sun affects our weather in many ways, it is not simply a function of the minute-by-minute radiation striking Earth during daytime on a clear day. Karoly is over-simplifying the complex and poorly understood interactions between solar variability and Earth's weather.

Previously we showed how the 11-year solar cycle affects some ENSO events. The coldest portion of the Little Ice Age (1645–1715) coincides with the Maunder Solar Minimum, when sunspots became exceedingly rare. As discussed in Chapter 4, sunspots reflect solar activity; more sunspots, more solar output; fewer sunspots, less solar output.

When Karoly says "changes in solar radiation" he is referring to variations in TSI or the Sun's total *radiation* output. Further, he is talking about the variation over days, months, and up to a few decades. TSI varies little over the short term, roughly ±0.2 W/m^2, over an eleven-year solar cycle, after correcting to the amount that strikes Earth's surface. The correction accounts for the radiation reflected by the atmosphere and surface (~30% reflected) and for Earth being a sphere (divide by 4).

We have seen that the IPCC believes that humans have caused 2.3 W/m^2 of change at the Earth's surface since 1750, a much larger effect because the solar cycle is up and down approximately every eleven years. But the solar cycle is only a short-term change in total radiation output. There are other, generally longer term, changes in the Sun's magnetic field and other solar forces as discussed in Chapter 4.

Unlike the solar cycle, small continuous changes in solar radiation output, the solar magnetic field, or in the Sun's particle output (the solar wind) can integrate over time. These changes might be small, but, if they continuously increase or decrease the thermal energy stored in the oceans, they can eventually have a large impact on the climate. Oceans cover over 70% of Earth's surface, and as they accumulate energy, they move it around in currents that alternately release or store the energy in a pseudo-cyclical pattern, or oscillation. The oscillations operate with various delays that ultimately have a global effect on the weather.

Besides the ENSO oscillation, there are the well-known PDO and AMO. There are even longer known oscillations, such as the 2,450-year Bray cycle and the 1,000-year Eddy cycle. These, and more, are detailed in (May, 2016k). Another good reference is (Vinós, 2016).

The exact way the Sun influences Earth's climate is not known, but known changes in solar activity, as documented in sunspot records, coincide with historical changes in Earth's climate. Nir Shaviv has analyzed the change in ocean-heat content over the past 50 years via several methods and found that the radiative forcing associated with solar-cycle TSI variations is five to seven times higher than can be accounted for by the change in TSI alone. There must be an amplifier involved that increases the solar effect. He suspects, but cannot prove, that the amplifier is related to cosmic-ray flux in the atmosphere modulated by changes in the solar magnetic field (Shaviv, 2008).

Shaviv's results are consistent with previously calculated radiative imbalances due to cosmic-ray attenuation by the solar magnetic field. When cosmic rays are attenuated, lower cloud cover decreases and surface temperatures rise, when cosmic rays are allowed in, lower cloud cover increases and surface temperatures fall (Marsh & Svensmark, 2000).

Further, the IPCC assumes our Sun varies less over longer periods of time than similar stars. This is illustrated in Figure 26, in Chapter 4.

Less oxygen in the air

Figure 47 also tells us that there is slightly less oxygen in the atmosphere today because of burning more fossil fuels. This has been measured and it is logical. It tells us nothing about the human influence on global warming, but it is true.

Background: Cooling upper atmosphere

The cooling upper atmosphere, or the so-called anthropogenic fingerprint, was introduced in Chapter 4. The fingerprint issue, and the politics around it, are further detailed in our earlier book, *Politics and Climate Change: A History* (May, 2020c, pp. 233-235). Here we only need to emphasize that there are many reasons why the upper atmosphere may be cooling while the lower atmosphere warms. It is true that stratospheric cooling is predicted by the climate models and by both Happer and Karoly. But anthropogenic CO_2 is not the only possible cause. Even if additional anthropogenic CO_2 is causing stratospheric cooling, it tells us nothing about how much of tropospheric warming is due to human emissions. Further, as we discussed in Chapters 4 and 5, the climate modelers have had a very tough time predicting upper atmospheric temperatures, especially in the critical tropical troposphere. Comparing Figures 18 and 19 in Chapter 4 illustrates this point very well.

Figure 47 Summary

Figure 47 is a useful summary of Karoly and Tamblyn's evidence that humans may be causing global warming. However, all the evidence is qualitative and ambiguous about the human impact on climate. Of the 13 items listed, five result from the burning of fossil fuels and don't speak to warming or climate change. Two items—less heat going to space and a rising tropopause—are an obvious result of surface warming, regardless of the cause.

The Barnett and colleagues 2005 pattern-of-ocean-warming study suffers from the same problems as all climate-model studies. If you assume that the Sun and the ocean's impact on the climate are invariant, then assume that the only other factors in global warming are humans and volcanism, you bake in the answer before you start.

Once these Figure 47 items and the unsupported shrinking upper atmosphere idea are eliminated, only four remain: cooling upper atmosphere; winter and nights warming more than summers and days; less heat escaping to space; and more heat returning to Earth. All four could possibly result from human-greenhouse-gas emissions, but all could be caused in part, or entirely, by natural forces. None of the four are quantitative. Happer believes we are adding GHGs to the atmosphere and that they should cause some warming, he is just unconvinced the warming or the gases are dangerous.

One problem with the remaining items is that all of them might only reflect changes in cloud cover. Clouds and cloud cover cannot currently be modeled, they must be assumed or "parameterized." The amount and type of clouds in the atmosphere can strongly affect climate, and we still do not know if greater cloud cover increases or decreases warming. The relationship between cloud cover and climate is complex and unpredictable. Clouds are so important and powerful in global warming that small changes can completely nullify the implicit assumptions in all the items listed in Figure 47. We look at clouds more closely in the next section.

Cloud Cover and Climate Change

Happer points out in his Major Statement, that much of Earth is covered in clouds and that they are largely responsible for its albedo, which reflects about 30% of sunlight back into space. He adds that while the effects of clouds are comparable to the effects of GHGs, they are much harder to model. Clouds work to cool the Earth when they reflect sunlight during the day, but at night they work to warm the surface by capturing surface radiation, warming the air around them, and radiating some of the captured energy back to the ground. Happer adds that low clouds radiate energy, in all directions, at a temperature near the surface temperature, in addition to reflecting sunlight during the day, so they are probably net cooling. Bodies at higher temperatures radiate more energy than bodies at lower temperatures.

Upper atmospheric cirrus clouds are very cool, relative to the surface, so they have a warming effect. Their cold temperature does not allow them to radiate as much to space as they absorb from below.

Karoly doesn't discuss the effect of clouds and Tamblyn has little to say. Tamblyn does mention that changes in cloud cover might affect the strength of the greenhouse effect, which agrees with the conclusions of the IPCC in AR6. In AR6, we see: "It is well known that most of the model spread in ECS arises from cloud feedbacks, and particularly the response of low-level clouds." (IPCC, 2021, p. 7-113).

Background: Clouds

Internal variability, ocean oscillations, and changes in cloud cover have a large effect on our weather, and all are likely related, at least in part, to variations in solar activity. Additional cloud cover cools the surface during the day and warms it at night. Small changes in cloud cover can have a large effect. Cloud feedback is, by far, the largest source of uncertainty in climate models.[119] The net effect of clouds on surface temperature can be negative or positive, depending upon the vertical distribution of the clouds, the cloud types, and other factors. Other factors include latitude and season, outside the tropics clouds warm more at night and during the winter, when they also tend to be more abundant. But surface temperature changes can also affect cloud cover. Thus, clouds are both a forcing and a feedback with regard to global surface temperature changes.

A connection between the number of cosmic rays hitting Earth's atmosphere and the 11-year solar cycle has been established for many years (Svensmark, 1998). Henrik Svensmark has proposed a more controversial connection between the number of cosmic rays striking the atmosphere and

---

[119] (Ceppi, Brient, Zelinka, & Hartmann, 2017)

total cloud cover and has gathered convincing evidence of the connection (Svensmark, 2019). But, regardless of the relationship between cloud cover, solar activity, and cosmic rays, clouds are a huge factor in climate change.

Recent studies of climate-model output show that nearly all climate models treat increasing cloud cover as a positive feedback. That is, they are constructed in such a way that increased surface temperatures, increase cloud cover, and this leads to increased warming. However, there are some climate models that produce a negative (cooling) cloud feedback. The total spread of modeled cloud feedback is from -0.13 $W/m^2/°C$ to 1.24 $W/m^2/°C$.[120] AR6 presents a similar range, from -0.1 to 0.94 $W/m^2/°C$ (IPCC, 2021, p. TS-60). The IPCC states that "Uncertainty in the sign and magnitude of the cloud feedback is due primarily to continuing uncertainty in the impact of warming on low clouds." (IPCC, 2013, p. 16). AR6 states that "clouds remain the largest contributor to overall uncertainty in climate feedbacks (*high confidence)*" (IPCC, 2021, pp. TS-59). Real-world observations from the Clouds and the Earth's Radiant Energy System (CERES) satellites (NASA, 2021) suggest that overall, clouds work to cool Earth, regardless of their status as a surface warming feedback (Eschenbach, 2017).

Clouds rapidly adjust in response to surface warming and thus to CO_2 forcing of warming. Paulo Ceppi and colleagues[121] describe, through a series of digital experiments, how clouds are modeled in the AR5/CMIP5 climate models. The equation below is at the heart of their experiments:

$$N = F + \lambda \cdot \Delta T \ldots \text{Eq. 1}$$

N is the energy-flux imbalance (incoming-outgoing) at the top of the atmosphere (TOA), F is a positive forcing, in W/m^2 due to a sudden increase in greenhouse gases, λ is the cloud feedback in $W/m^2/°C$, and ΔT is the total global temperature change in °C required to regain equilibrium, or an N of zero. A hypothetical situation used in the paper to calculate F was an instant quadrupling of the CO_2 relative to pre-industrial conditions. Their feedback numbers cannot be duplicated with real-world data due to the implausible scenario. Here are two more forms of the same equation for reference.

$$\Delta T = (N-F)/\lambda \quad \text{or} \quad \lambda = (N-F)/\Delta T \ldots \text{Eqs. 2 \& 3}$$

We define the top of the atmosphere as the orbits of the CERES satellites. N is positive if the downward force is larger (warming) and negative if the outgoing radiation is larger (cooling). The Earth is at equilibrium if the feedback, N and F are zero. Like N, positive feedback (λ) leads to warming.

---

[120] Ibid.

[121] (Ceppi, Brient, Zelinka, & Hartmann, 2017)

The higher the feedback, the greater the warming. If the feedback is negative; cooling, or slower warming is the result.

After an initial forcing, if nothing else changes, N will eventually return to zero and the surface temperature will change. The temperature change is determined by equation 2, which when N=0, reduces to $\Delta T = -F/\lambda$. To regain N=0, or a new equilibrium, at a new temperature, λ must be negative, if F is positive. That means the outgoing radiation can increase to a new level and everything stabilizes. If λ is positive, then more warming is expected, and it goes on and on. This is the fear of endless global warming in a nutshell. Are the total natural feedbacks to additional CO_2 positive or negative and how large are they?

The Cloud-Radiative Effect (CRE) is the difference between the clear-sky and all-sky radiative flux at the satellite. Figure 48 is a map of the average TOA CRE imbalance at the CERES satellites from 2001 through 2019.

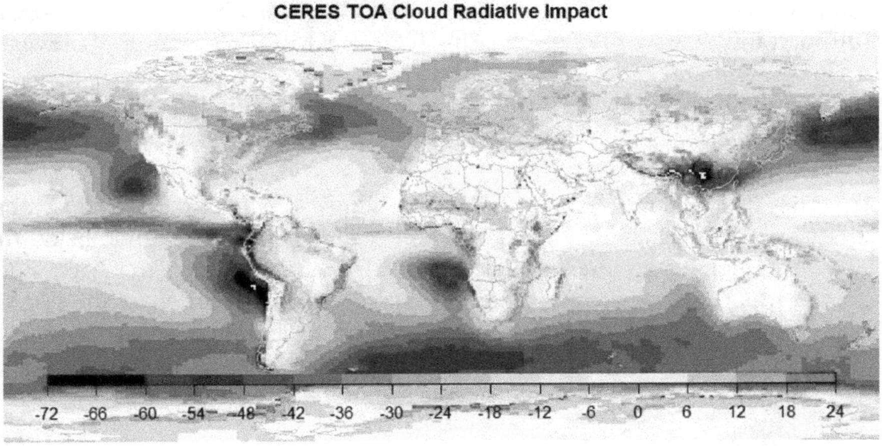

CERES TOA Cloud Radiative Impact

Figure 48. CRE in W/m^2. Negative values (black, gray and blues) are cooling. Data from NASA.

Clouds reflect incoming solar shortwave radiation (SW), so more SW travels up to the satellite in the presence of clouds. On average, the difference is about −45 W/m^2. The number is negative because more radiation is leaving Earth—a cooling effect. Clouds also block some outgoing longwave-infrared radiation (OLR) emitted by Earth's surface, on average about 27 W/m^2, a positive number. A positive number is energy retained by Earth and not reaching the satellite—a warming influence. The difference is −18 W/m^2, which means overall: clouds cool the Earth.

One would think that the more clouds, the more rapidly Earth would cool, but it is not that simple. As Happer mentions in his Major Statement some clouds, especially low-level clouds and cumulus clouds, tend to reflect more energy during the day than they trap at night. High level clouds, like cirrus, tend to allow a lot of solar SW through and they emit little energy to space because they are so cold. But they absorb a lot of upwelling OLR, thus they have a net warming effect. Cloud type, location, distribution, and altitude matter.

Figure 48 is colored blue where clouds are net cooling and in pink, yellow, and red when the clouds are net positive. The effect is negative (or cooling) almost everywhere. The exceptions are deserts and polar land regions—areas where clouds tend to trap surface and lower atmosphere infrared radiation and simultaneously allow solar SW through to the surface and/or emit very little radiation to space. The latter combination creates a positive CRE and warming.

Figure 49 shows the average monthly cloud cover as measured by the CERES satellites from 2001 through 2019. White regions represent areas of greatest cloud cover and darkest areas represent the least. The Intertropical Convergence Zone—where the trade winds of the Northern Hemisphere and Southern Hemisphere converge—is marked by a nearly horizontal light streak through the center of the figure. The ITCZ is where evaporation of seawater is maximum, and the air is most humid. The rapidly rising humid air brings frequent rain and thunderstorms. It is almost always cloudy in the ITCZ, and the cooling effect of the clouds is very high, as you can see in northwest South America in Figure 48.

The regions of maximum cloud cooling—those displaying the most negative CRE values in Figure 48—are in the small white spots inside the darkest regions in southern China and off the coast of Peru. The CRE values in the spots are so negative (extreme cooling) they are off scale and appear white. Figure 48 correlates reasonably well with the cloud fraction in Figure 49, except in the polar regions.

CERES Cloud Fraction (%)

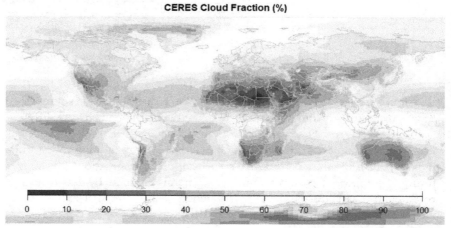

Figure 49. The CERES cloud fraction, monthly average from 2001 through 2019. Source NASA.

Figure 50 is NASA's blue marble photograph with ice and clouds shown on a Mercator projection. Outside of the poles, notice the similarity with Figure 49.

Figure 50. The NASA Blue Marble, shown with a Mercator Projection.

Figure 51 shows the same monthly CERES data used in Figure 48 in light gray, and the yearly averages as a heavy black line. As Norman Loeb and colleagues[122] explain, Earth's continuous energy imbalance is so tiny that it is only 0.15% of the total incoming and outgoing radiation.

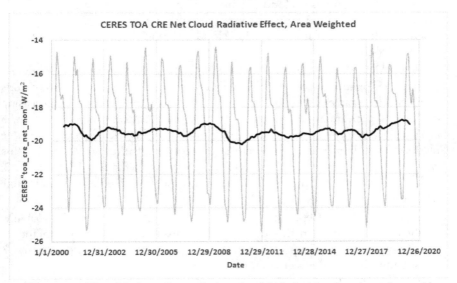

Figure 51. The global average radiation cloud imbalance as yearly and monthly averages, compare to Figure 48. Data: NASA.

---

[122] NASA Langley Research Center

AR6 estimates that the imbalance from 1971 to 2018 is 0.79 W/m² (0.52 to 1.06), or about 0.02 W/m²/year (IPCC, 2021, p. 7-6). The net incoming-outgoing imbalance is the difference between two large numbers and below the uncertainty in the satellite measurements.

Due to the coarse resolution of the CERES instrument, there are a lot of missing grid cells in the one-by-one degree latitude and longitude grid used to make the maps in Figures 48 and 49. To get around these problems Loeb and colleagues use a complex algorithm called "Energy Balanced and Filled," or EBAF, to fill in missing values and adjust the SW and OLR TOA fluxes, within their uncertainty ranges. These adjustments are meant to remove inconsistencies between the global net TOA flux and the heat storage in the earth-atmosphere system (Loeb, et al., 2018). EBAF does not improve the accuracy, but it makes the maps and graphs look better.

The CRE values in Figure 51, vary widely from month to month, and the net yearly average value throughout the period from 2001 to 2019 is –19.1 W/m², which is very close to Ceppi and colleagues' value of –18 W/m². There is no apparent long-term trend.

Cloud cover is graphed in Figure 52 in orange, using the CERES data mapped in Figure 49. There is an apparent inverse relationship between cloud fraction and the cloud radiative effect (CRE) plotted in black. Both the cloud radiative effect and the cloud fraction are plotted as area-weighted yearly averages. As the cloud fraction decreases, the cooling effect becomes less negative, that is warming. As the cloud fraction increases, the opposite occurs—cooling.

Both AR5 and AR6 suggest that the overall effect of increasing cloudiness is warming, or more precisely, they suggest that as the surface warms, clouds increase the warming rate. Figure 52, suggests the opposite, increasing cloudiness appears to make the CRE radiation balance more negative, a cooling effect. This is not exactly the same as saying clouds are a negative feedback to rising surface temperatures, but it suggests they are. Combine this with the fact that overall clouds cool Earth, as seen in Figures 48, 51, and 52, and the IPCC assumption that clouds are a positive feedback is definitely debatable.

Figure 53 shows Ceppi and colleagues' modeled cloud feedback globally. In their work, just as in this book, a more positive feedback parameter (λ) implies warming. Working with models and model output only, they derive λ by isolating the cloud effect from other feedbacks.

Ceppi and colleagues note that the CRE response to temperature changes is not the same as the feedback value they compute, because CRE also responds to changes in albedo or water vapor. Thus, the CRE response to surface temperature changes is about 0.3 W/m² less than their feedback value. The units in Figure 53 are W/m²/°C, of modeled surface warming or cooling due to clouds over the time it takes to reach equilibrium. The cloud-feedback parameter shown in Figure 53 is positive, and according to Ceppi and colleagues, cloud feedback tends to be positive suggesting that clouds respond to warming by causing more warming.

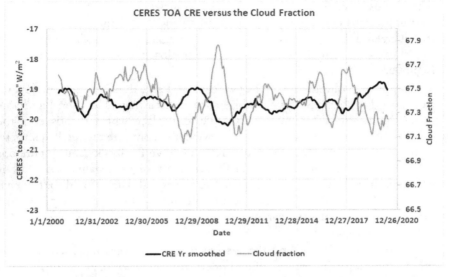

Figure 52. *The CERES average cloud fraction in percent, from Figure 49, is shown in orange and compared to the CERES yearly CRE from Figure 51 in black. Data source: NASA.*

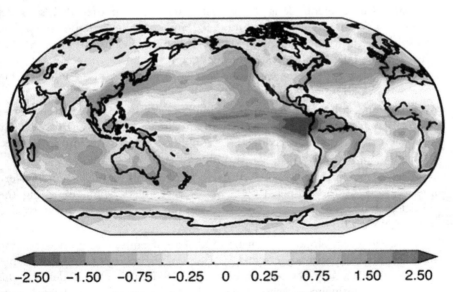

Figure 53. *Ceppi and colleagues' map of the modeled cloud feedback parameter.*

As Willis Eschenbach points out Figure 53 is a plot of model output and Figure 48 is data (Eschenbach, 2017). The data in Figure 48 is massaged and close to the edge in terms of uncertainty, but it is data. It is not strictly speaking, a map of the cloud feedback to surface warming, but it does show that clouds overall work to cool Earth's surface. The difference, that is warming versus cooling, between Figures 53 and 48 is large and very difficult to understand. It is possible that overall cloudiness has a cooling effect, as in Figure 48, and is simultaneously a positive feedback to surface warming, as in Figure 53, but it seems unlikely.

Ceppi and colleagues are lucky we cannot derive their feedback parameter from real data, because if we could, I suspect their map, shown in Figure 53, would look very different. For example, one place where clouds cool the surface the most is offshore of Peru. If we could compute the feedback there, would it still show as an area with a large positive feedback, as in Figure 53? In southern China, a relatively small region of slight negative feedback is modeled there, but nothing like we see in the actual data. Would that remain so small? The cloudy ITCZ is shown as remarkably hot in Figure 53. Would that stay the same if data were used in the calculations? Most of the ITCZ is quite dark in Figure 48.

In AR6, the IPCC admit the eastern Pacific is always modeled with too much warming. Sometimes the models predict two degrees more than observations, as discussed earlier in this chapter (IPCC, 2021, p. 7-88). Ceppi's plot seems to explain part of their problem.

The idea that clouds are a net-positive (warming) feedback, makes no sense. The worst of it is that nearly every model produces a positive cloud feedback. Cloud feedback is the largest component of the IPCC's model derived ECS. Clouds cannot be modeled and must be parameterized (the modeling term for "assumed"). Steven Koonin explains in *Unsettled* that scientists from the Max Planck Institute tuned their climate model by targeting an ECS of about 3°C. This was done by adjusting their cloud feedbacks. He remarks "talk about cooking the books." (Koonin, 2021, p. 93).

Ceppi and colleagues report that cloud feedback is, "by far, the largest source of inter-model spread in equilibrium climate sensitivity (ECS)." They also point out that cloud feedback is strongly correlated with model derived ECS and supply us with the data. It is plotted in Figure 54.

Oops! The relationship between cloud feedback and modeled ECS is very linear. It seems the modeled ECS *can be* controlled by manipulating the modeled cloud feedback. This only works in the climate model world; we live in the real world.

As we've seen, clouds cannot be modeled currently, and models assume that clouds have a warming effect. CERES satellite data says clouds have a net cooling effect, a *large* net cooling effect of -18 W/m². The models say the entire human influence on climate since the beginning of the industrial era is 2.3 (1.1 to 3.3) W/m² (IPCC, 2013, p. 661), which puts the cloud impact of -18 W/m² into perspective. Notice the year-to-year variability in Figure 51 is larger than one W/m², and it only covers 20 years. How much of the ECS from models is due to their assumption that clouds are net warming? How much is due to their assumption that ECS is 3 W/m² (IPCC, 2021, pp. TS-58)?

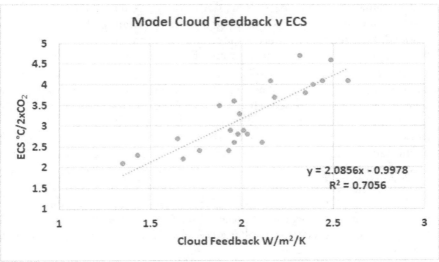

Figure 54. This is the modeled cloud feedback (x axis) from Figure 53, plotted versus model derived ECS. Data source (Ceppi, Brient, Zelinka, & Hartmann, 2017).

The albedo of specific clouds varies, from 10% to 80%, depending upon type, location, altitude, and the local time of day. According to NASA, Earth's total albedo varies ±0.4% from year to year, without a specific trend (NASA, 2021). In one year, it can cause a swing of over 0.8 W/m² in solar forcing, this is 34% of the IPCC estimate of the total human influence on climate since 1750 cited above. We are not sure if the albedo is increasing or decreasing with time—although the CERES satellite data suggests it has decreased slightly since 2000 (Koutsoyiannis, 2021). At the same time, Demetris Koutsoyiannis's analysis of MERRA-2 satellite data shows both total and cloud albedo increasing since 1980. The relationship between cloud cover and albedo is an incredibly complex topic, that will not be solved soon. Neither will the relationship between clouds and global warming.

Summary of the evidence

The very small impact that the IPCC claims humans are having on Earth's climate must be compared to natural variability. Natural variability has not been measured or estimated to everyone's satisfaction, and as a result we don't know how much warming is due to natural forces and how much is due to humans. The IPCC shrugs this off and claims that overall, natural forces sum to zero over climatic periods of time. The Little Ice Age, the Medieval Warm Period, and the Holocene Climatic Optimum belie this assumption.

Karoly acknowledges that there is a lot of natural variability from year to year and decade to decade. He points to the cooling from 1945 to 1975 and from 2000 to 2012 as examples. He believes, as do most climate scientists, that these anomalous cooling periods are due to ocean oscillations, like ENSO. He does not believe that these cooling anomalies mean global warming has stopped, he thinks that it means more thermal energy is being absorbed into the oceans and less is going to warming the surface (Karoly, 2021a, p. 16).

The cooling upper atmosphere (stratosphere) mentioned in Figure 47, could be due to more CO_2, because CO_2 radiates a lot of infrared radiation to space from the stratosphere. It could also be due to increased shielding of thermal radiation from below due to additional IR GHG absorption in the troposphere or by clouds. It could also be due to less ozone. But, regardless of the cause, as the troposphere has warmed, the stratosphere has cooled, at least since 1979. For a discussion of the significance of the correlation, displays of the data, and additional references see (Munshi 2018).

The stratospheric cooling is not related to the tropospheric hot spot; that is a separate issue. Jamal Munshi found that there is no statistically significant evidence that observed stratospheric cooling is related to rising atmospheric CO_2 or tropospheric warming.

The evidence presented in Figure 47 that humans are causing global warming is not conclusive or convincing. The evidence Karoly presents that the Sun has little impact on climate only applies to one narrow portion of solar variability: TSI. It does not address other ways the Sun affects our climate including variations in the solar wind, the strength of the solar magnetic-field, and variations in solar UV radiation. All vary more than TSI.

How much the Sun and internal variability affect our climate versus human influences is unknown. The essence of the debate is not if CO_2 warms the lower atmosphere, but by how much. As we have seen, Karoly and the IPCC ignore warming due to nature. Once this is done, all warming can be attributed to humans. It is circular reasoning. More detail on Figure 47 and the supporting references can be seen at skepticalscience.com (Cook J. , 2010).

Chapter 10: Are CO$_2$ and Global Warming good or bad?

TheBestSchools.org:

> "The IPCC's official position may be summarized as making four claims: global warming is a well-established fact; it is anthropogenic; it is a major problem for humanity; and concerted global governmental action is required to combat it." (Happer, 2021a, p. 16)

The quote above is from the first question put to Dr. Happer. TheBestSchools asked both Happer and Karoly to discuss the statement. Following are their responses.

Karoly and Happer: What needs to be done?

Dr. Happer:

> "There is no scientific evidence that global greenhouse gas emissions will have a harmful effect on climate. Quite the contrary, there is very good evidence that the modest increase in atmospheric CO$_2$ since the start of the Industrial Age has already been good for the Earth and that more will be better." (Happer, 2021c, p. 2)

Dr. Karoly:

> "Science has established that it is virtually certain that increases of atmospheric CO$_2$ due to burning of fossil fuels will cause climate change that will have substantial adverse impacts on humanity and on natural systems. Therefore, immediate, stringent measures to suppress the burning of fossil fuels are both justified and necessary." (Karoly, 2021b, p. 28)

Dr. Happer, also writes:

> "More CO$_2$ in the atmosphere will be good for life on planet Earth. Few realize that the world has been in a CO$_2$ famine for millions of years—a long time for us, but a passing moment in geological history." (Happer, 2021b, p. 32)

Dr. Happer again:

> "If increasing CO_2 causes very large warming, harm can indeed be done. But most studies suggest that warmings of up to 2 K [°C] will be good for the planet, extending growing seasons, cutting winter heating bills, etc." (Happer, 2021b, p. 26).

Is Global Warming a bad thing?

Happer states his opinion that the IPCC has grossly exaggerated the warming effect of CO_2. The IPCC believe that doubling CO_2 will likely cause about three degrees Celsius of warming (the ECS). Happer believes that it will cause roughly one degree of warming (Happer, 2021a, p. 20). Three degrees might be a "marginal" problem he says, but one degree will be beneficial. It is worth pointing out that the IPCC AR5 says ECS lies between 1.5 and 4.5 degrees and has been unable to narrow this range in over 40 years. The AR6 *very likely* range is two to five, but the uncertainty is still three degrees (IPCC, 2021, pp. TS-39). Empirical and theoretical estimates range from 0.4 degrees to over six degrees. The actual value is important, but hard to nail down.

A rise of only one degree is not alarming, a point generally agreed upon by both sides. The question revolves around the feedbacks related to this increase, are they positive, increasing the warming, or negative, decreasing it? Decades of research, billions of dollars, and thousands of scientists have not been able to determine the answer. Climate models go both ways and are of no help. Glenn Tamblyn writes:

> "Global Warming is not good. Inaction to prevent global warming is not good. Technological progress is rapidly removing any impediments to taking action. What possible reason could remain for not acting?" (Tamblyn, 2021b, p. 24)

Karoly writes that we are already seeing adverse impacts from global warming and that these will continue in the future.

> "Global warming has led to increases in hot extremes and heatwaves, affecting human health and leading to crop and animal losses, as well as increases in the occurrence and intensity of wildfires in some regions.
>
> Increases in global temperature have led to global sea level rise, flooding coastal areas and causing coastal erosion and pollution of coastal freshwater systems with seawater." (Karoly, 2021b, p. 21)

Heat extremes may have increased, it is hard to say. Some records suggest they were worse in the United States in the 1930s, but certainly winters are warmer now and so are nights (IPCC, 2013-WGII_TS, p. 52). State temperature all-time highs, in the United States are from long ago, most of them are from before the 1940s.[123] Also, according to NASA, the global burned area has declined 25% from 2003 to 2019.[124] U.S. Forest Service data suggests wildfires were much worse in the 1930s than today.[125]

Karoly continues:

> "Expected sea level rise by the end of this century for even the smallest projected global warming will lead to the annual flooding of many hundreds of millions of people and the complete loss of some low-lying island countries.
>
> "One of the other major impacts of climate change due to increasing carbon dioxide concentrations is the increase in carbon dioxide dissolved in the oceans. As shown below, the dissolved carbon dioxide in the upper waters of the ocean has increased in parallel with the increase in atmospheric concentration. As the oceans absorb more carbon dioxide, they become less basic (or more acidic), with a higher concentration of carbonic acid. This can be seen in the decrease in pH of ocean water by about 0.1 units over the last 30 years." (Karoly, 2021b, pp. 21-23)

We discuss ocean pH, rising sea level and extreme weather later in this chapter. Happer writes about imaginary climate change dangers (bogeymen):

> "One of the bogeymen is that more CO_2 will lead to, and already has led to, more extreme weather, including tornadoes, hurricanes, droughts, floods, blizzards, or snowless winters. But as you can see from Fig. [55], the world has continued to produce extreme events at the same rate it always has, both long before and after there was much increase of CO_2 in the atmosphere. In short, extreme weather is not increasing." (Happer, 2021b, p. 36)

Happer's figure is our Figure 55. It is a bit difficult to interpret, so we will show some easier- to-interpret figures from our first book, *Climate Catastrophe! Science or Science Fiction?* (May, 2018).

---

[123] https://nationalweathermuseum.com/all-time-state-high-temperatures/
[124] https://earthobservatory.nasa.gov/images/145421/building-a-long-term-record-of-fire
[125] https://climateataglance.com/climate-at-a-glance-u-s-wildfires/

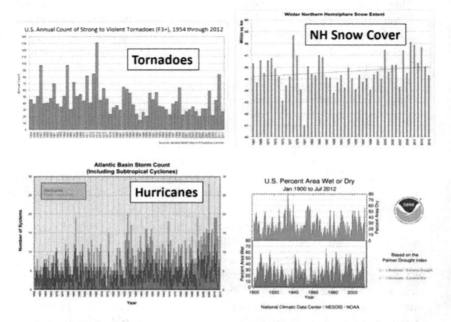

Figure 55. Happer's Figure 18. It shows tornados have declined, snow cover has increased and droughts and flooding in the U.S. have not changed. Major hurricanes (violet) and normal hurricanes (red) have not changed significantly. Source: NOAA.

Figure 56 plots 10-year sums of hurricane central-pressures subtracted from 1020 mbar. We subtract from normal sea level pressure, so larger numbers indicate more and stronger hurricanes. The graph clearly shows a decline in the number and strength of hurricanes over the past 170 years.

Figure 56 only applies to the U.S., which is a small portion of Earth's surface. Figure 57 shows the number of global weather and climate disasters from 2000 to 2021, as reported by EM-DAT in Belgium. EM-DAT only has reliable data since 2000. The data before 2000 is of poorer quality and only reflects an increase in reporting according to the EM-DAT data manager Regine Below, so it is not shown here.

So, the total number of weather and climate related disasters is declining, but that is only part of the story. U.S. weather-related disaster costs, as a percent of GDP, have also declined since 1990 (Pielke Jr., 2022b). Pielke Jr. has also shown that global disaster costs as a proportion of global GDP have declined. The global population grows every year and every year more of these people move to areas vulnerable to weather-related disasters, such as coastal

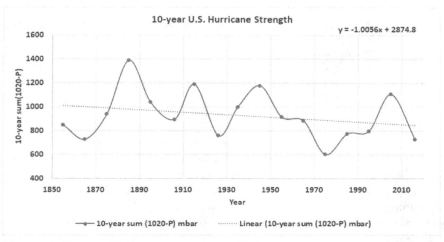

Figure 56. Ten-year U.S. hurricane strength. Higher numbers mean stronger and more hurricanes. Source: (May, 2018, p. 182)

areas, forests, mountain sides, etc. Further, people are more affluent every year, so they live in more expensive houses than in the past, yet the cost of weather-related disasters as a percentage of GDP—while higher in nominal terms—is less.

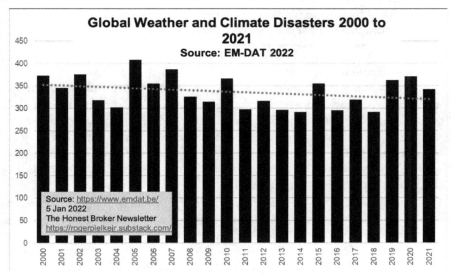

Figure 57. The number of global weather and climate disasters from 2000 to 2021. Source: (Pielke Jr., 2022). Data source: EM-DAT.

This means disasters have a smaller impact on our economic well-being and the trend is expected to continue. Humans are adapting to extreme weather better as they become more affluent. In part, this reflects better construction standards. The IPCC appears to agree, and we find this in AR5:

> "Economic losses due to extreme weather events have increased globally, mostly due to increase in wealth and exposure, with a possible influence of climate change (*low confidence* in attribution to climate change)." (IPCC, 2013-WGII_TS, p. 49).

The IPCC is clearly saying that the risks we face from extreme weather are due to our greater wealth and more exposure to these events. They have low confidence that it is due to [human-caused] climate change. We add the qualifier "human-caused" to the sentence because climate change is a meaningless phrase by itself, climate always changes. In fact, the phrase climate change is redundant, there would be no need to name climate if it never changed.

Karoly also writes:

> "Indeed, many studies have concluded that the impacts of climate change are not distributed uniformly. The impacts are greater for the poor, disadvantaged, and indigenous populations around the world, partly because they have greater exposure and are more vulnerable to the impacts of climate change and partly because they have fewer resources to adapt to or to avoid these impacts." (Karoly, 2021b, p. 21)

His statement is perfectly OK, as written. It is the underlying assumptions posed by it that are not established. To say that poor and disadvantaged people are more vulnerable to extreme weather is true. More affluent people build better houses and have the financial resources to rebuild them after weather disasters. The optimum answer is to lift the poor from poverty, increasing the cost of energy does not do this, it works in the opposite direction. The answer is not to impoverish the wealthy, the answer is to enrich the poor, so they become more resilient to change, whether man-made or natural.

As Happer says in his Major Statement, Earth has stubbornly refused to warm as quickly as predicted by the climate models, as we document in Chapter 5. Happer continues: "the climate establishment has invented a host of bogeymen—other supposed threats from more CO_2. It is almost comical to list them." (Happer, 2021b, p. 36). Presumably the bogeymen are intended to alarm the public, so the climate establishment can keep their jobs. Happer points out that they even changed the name from human-caused or anthropogenic global warming, a testable hypothesis (that was not testing well), to climate change. After all, who could possibly prove that climate does not change?

In Karl Popper's famous book, *Conjectures and Refutations, The Growth of Scientific Knowledge,* he defines pseudosciences, like astrology, Marxism, and psychoanalysis, as theories that are not falsifiable. Marxists and Freudian analysts saw confirmations of their theories everywhere. Every time a Marxist opened a newspaper, they found confirmation of their theory. Freudians saw their theory confirmed with every new patient. (Popper, 1962, p. 35).

To a climate alarmist, climate change is confirmed by every extreme weather event. They believe anthropogenic climate change is synonymous with climate change. Every weather event is caused by anthropogenic climate change. If it is too hot, too cold, too wet, or too dry, they are all caused by anthropogenic climate change. Thus, climate change is a pseudoscience because it is not falsifiable. Further, as Happer points out, the benefits of global warming or additional CO_2 are ignored or deemed irrelevant.

Anthropogenic Global Warming or AGW is a scientific hypothesis, it can be tested, and has been. Climate models are scientific tools, they make predictions that can be checked, we checked them in Chapter 5, and they failed (McKitrick & Christy, 2018). So, the goal posts were moved to climate change; something that cannot be checked.

Glenn Tamblyn has much to say about Happer's comments on extreme weather trends, or the lack thereof. He complains that Happer's Figure 18 (our Figure 55) focuses on the United States and the Atlantic. Therefore, we added Figure 57. He also dislikes Happer's plot of winter snowfall, Figure 55, upper right. He correctly points out that snowfall is also increasing in the fall, while decreasing in the spring and slightly declining over all (Tamblyn, 2021a, p. 19). This seems weak, considering David Viner and David Parker, at the Hadley Centre, famously predicted that children will not know what snow is by "a few years" after 2013.[126]

The IPCC published a report: *Managing the Risks of Extreme Events and Disasters to Advance Climate Change Adaptation* in 2018 (IPCC, 2018) in which they make several points relevant to this discussion:

- Attribution of single extreme events to anthropogenic climate change is challenging, page 7.
- Uncertainty in projections of future extremes is very large; page 113.
- Confidence in changes in extremes is low. Changes are local, not global; page 8.
- Economic, including insured, disaster losses from 2001 to 2006 were 1% of GDP for middle-income countries, 0.3% for low-income countries, and less than 0.1% for high-income countries; page 9.

---

[126] (Blair, 2013) and (Onians, 2000)

- Long-term trends in economic disaster losses adjusted for wealth and population increases have not been attributed to climate change, but a role for climate change has not been excluded; page 9.
- Most studies focus on cyclones, where confidence in observed trends and attribution of changes to human influence is low; page 9.

The report did not find a significant trend in tropical cyclone activity, tornados, or hail. Some regions had more drought, but other regions had fewer. The same with flooding. No global trends in any category of weather-related disasters were found. As weather extremes are always occurring somewhere, claiming that any of them are due to human activities is meaningless. Most of the report relies on *projections* of *possible* problems in the future, because attributing extreme weather events to humans is "challenging." In general, claims that human GHG emissions are somehow making weather extremes worse or more frequent have no basis in observations. The claims are based on projections made with unvalidated models.

Karoly writes that sea level rose about 20 cm (about 8 inches) from 1900 through 2016. He tells us this is due to the water in the oceans expanding as it warms and the melting of glaciers on land. Happer uses recent trends to project a sea level rise of about the same amount over the next century (Happer, 2021b, p. 37).

Tamblyn speculates that the rate of sea-level rise is accelerating, and that sea level might rise "several meters in decades" without offering any evidence or references (Tamblyn, 2021a, p. 23). The rate of sea-level rise varies from year to year and is so small and the complexities of measuring it in an ocean with waves over a meter high mean that the error in measurement is much greater than the rate of rise. All sea level measurements should be taken with a large grain of salt. As Kip Hansen points out, the NOAA specification for tide-gauge accuracy, for a monthly mean, is ±5 millimeters.[127]

The recent rates cited by Tamblyn vary from 1.7 to 3.4 mm/year. This cannot reasonably be called acceleration when the tools used to measure sea level cannot distinguish between the two values (Hansen K., 2020). Geological records suggest that Meltwater Pulse 1A (shown in Figure 58) saw sea level rising over 20 mm per year, ten times the observed rate today (Cronin, 1982). We do not know how fast sea level is rising with any precision and the rates observed are too small to matter.

The next bad thing is ocean "acidification." Happer writes the following in his Major Statement: "In biologically productive areas, photosynthesizing organisms remove so much CO_2 during the day that the pH can increase by

---

[127] More specifically, the 2013 NOAA tide-gauge accuracy requirements are 1 mm or better for tidal ranges of 5 meters or less, 3mm for 5-to-10-meter tidal ranges, and 5mm or better for tidal ranges greater than 10 meters (NOAA, 2013).

0.2 to 0.3 units, with similar decreases at night when respiring organisms return CO_2 to the water."

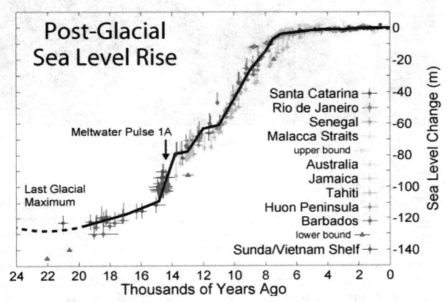

Figure 58. A sea-level reconstruction for the past 24,000 years by Robert Rohde.

Thus, Happer does not believe the estimated average decrease in pH of 0.1 unit cited by Karoly (Karoly, 2021b, p. 23) is significant. After all, if the daily local range of pH is over 0.4, how can this be a problem? Life thrives all the way from Florida Bay to Key West, as anyone who has snorkeled there can attest, yet the pH in this region varies from around 5.8 in the mangrove fringes in winter to 9.6 near Key West in the summer (Istvan, 2021). These extremes are driven by rainfall, the vegetation, and increased evaporation in the summer. It is safe to say that any organism that cannot handle a change in pH of 0.4 would have extirpated long ago.

Tamblyn tries to make the case that it is not the change in pH that matters, but the speed of the change. But nature tells us that rapid pH changes over a much larger range than he suggests are common on a seasonal basis.

Tamblyn's next scary pH story is about the damage a lower pH might have on shelled creatures, like pteropods, microscopic ocean creatures made from aragonite, a softer version of calcite. He presents our Figure 59 (Tamblyn, 2021a, p. 27).

It turns out Tamblyn fell for a clever piece of propaganda. He thinks the images in Figure 59 are before and after pictures of pteropods (also called sea butterflies) in normal seawater and seawater undersaturated with carbonate due to the addition of CO_2.

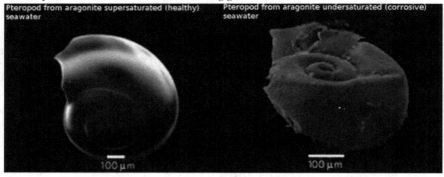

Figure 59. Tamblyn's illustration of a "healthy" pteropod, on the left, and one in "corrosive" sea water, on the right. Source: (Tamblyn, 2021a, p. 27)

Happer points out that pteropod shells are so fragile that few are preserved as fossils, but the fossils we do see show that they have existed in the oceans for many millions of years, back to when CO_2 concentrations were many times higher than today. Pteropods evolved in the Cretaceous geological period, more than 65 Ma, when CO_2 concentrations were between 1,000 and 2,000 ppm, some three to five times greater than what they are today (Peijnenburg, et al., 2020). Katja Peijnenburg and colleagues showed that, in the geologic past, all organisms with aragonite shells survived much warmer and higher CO_2 conditions than today or what we expect in the distant future.

Happer also points us to blog posts by Jim Steele.[128] The posts expose NOAA's erroneous hypothesis that additional CO_2 is causing carbonate to be undersaturated at the ocean surface and keeping animals from forming proper shells. When CO_2 is dissolved at the ocean surface, it does initially reduce pH, but is then taken up by plants through photosynthesis, which raises the pH. It is also incorporated into animal shells that sink to the ocean floor upon their deaths (many only live a few days), removing the carbon from the photic zone. Experiments have found that often CO_2 is a limiting nutrient, meaning that adding it to the ocean increases plant growth and abundance. The climate models do not capture this complexity, they simply model CO_2 as being absorbed by the oceans and then spread throughout the shallow ocean waters.

Background: Pteropods and CO_2

Far from being harmed by lower pH, sea butterflies regularly migrate from near the ocean surface, where the pH is typically about 8.1, to depths of one-hundred meters or more where the pH is around 7.6. They dive to the deeper waters nearly every winter in the higher latitudes.

All calcifying organisms have a protective organic layer that minimizes their sensitivity to changing pH. They also isolate their calcifying chambers from

---

[128] (Steele, 2017) and (Steele, 2017b)

ambient water conditions. The protective shell coating is called a periostraca. It protects the shell from corrosion if it remains intact. When the periostracum becomes damaged by predators or other incidents, pH—if it is low enough—can have an effect. There also exist a whole range of predatory organisms like bacteria, sponges and worms that can penetrate the periostracum and attack the pteropod.

These protective features allow shelled organisms to thrive in very low pH waters on the ocean floor and in lakes. In these areas the pH can drop to 6, and the water becomes truly acidic.

Both pteropods shown in Figure 59 are dead, and they may both be from the sea floor, hard to say. But the sea floor is commonly an environment with low enough pH that dissolution will occur. If any chemical etching occurs on the shell in shallow water, it will be due to upwelling deep water containing corrosive compounds. Shallow water is heavily buffered and kept in balance by biological activity. A living pteropod with full protective organic materials is shown in Figure 60.

James Orr and colleagues examined 14 living pteropods placed in ocean water with conditions they expected to exist in the far future in the Southern Ocean—conditions under the now debunked business-as-usual CO_2-emissions scenario, RCP8.5. All the pteropods survived, although a few exhibited some chemical etching on their shells.[129] It does not appear that a slight decrease in ocean pH, aka "ocean acidification," is a danger to shelled organisms, now or in the future.

Figure 60. A living pteropod.

---

[129] (Orr, Fabry, & Aumont, 2005)

There are many other claims about the dangers of ocean acidification we could discuss, but none of them bear close examination. In summary, the claims that extreme weather or a lower pH may cause problems are not based on any observations or mechanisms that can be documented today. They are based upon speculation about what might happen in the future. Worse the speculation is derived from unvalidated and discredited computer models.

Timothy Clark and colleagues investigated some published claims that ocean acidification was affecting fish behavior and could not replicate the results of previous studies.[130] They recommended that the Australian Research Council, the U.S. National Science Foundation, and the U.S. National Institutes of Health investigate possible fraud in 22 papers (Enserink, 2021). Many dangerous-ocean-acidification papers are in question.

Is Global Warming a good thing?

Karoly acknowledges that there are some possible benefits from higher CO_2 and warming, but these benefits apply only for "moderate levels of global warming." Thus, the magnitude and speed of expected warming is important. He writes:

> "The increase in carbon dioxide concentrations in the atmosphere has some potential benefits for plants because carbon dioxide is essential for photosynthesis. Plants grown in an atmosphere with higher carbon dioxide have faster growth rates and lower water use, assuming there are no other limits on growth." (Karoly, 2021b, p. 22)

Happer writes:

> "I believe that more CO_2 is good for the world, that the world has been in a CO_2 famine for many tens of millions of years and that one or two thousand ppm would be ideal for the biosphere. I am baffled at hysterical attempts to drive CO_2 levels below 350 ppm, or some other value, apparently chosen by Kabbalah numerology, not science." (Happer, 2021a, p. 42)

Many studies have been done to measure the impact of higher CO_2 on plant growth, Happer cites a paper from 2010 by Laci Gerhart and Joy Ward of the University of Kansas, in which the authors point out that during the Last Glacial Maximum, atmospheric CO_2 dropped to between 150 and 190 ppm, probably the lowest CO_2 levels since plants first evolved on Earth (Gerhart & Ward, 2010).

---

[130] (Clark, Raby, & Roche, 2020)

Vascular plants evolved during, or just preceding, the Devonian Period (419 million years ago). At that time, the atmospheric CO_2 concentration was 3,000 to 3,600 ppm. It has trended downward ever since, reaching a low of about 150 ppm during the Pleistocene Epoch glacial maxima. Between 18,000 and 15,000 years ago, CO_2 was between 180 and 220 ppm. Levels that low are highly stressful for C3 plants, which are 85% of all plant species. At that CO_2 concentration, plant productivity is greatly reduced, and some plants may fail to reproduce (Ward, 2005).

"It is clear that modern C3 plant genotypes grown at low [CO_2] (180–200 ppm) exhibit severe reductions in photosynthesis, survival, growth, and reproduction, suggesting that reduced [CO_2] during glacial periods may have induced carbon limitations that would have been highly stressful on C3 plants" (Gerhart & Ward, 2010)

Generally, additional CO_2 increases the productivity of C3 plants, at least, to a CO_2 concentration of 1,200 ppm, which is about three times the atmospheric CO_2 we have today. Above that, they do well up to 10,000 ppm (25 times today), and beyond that, their seed yields begin to decline (Ward, 2005).

The growth of 14-day old C3 plants, under various CO_2 levels is shown in Figure 61 (Gerhart & Ward, 2010). Happer includes this figure in his Major Statement on page 33. The growth under glacial conditions (150 ppm) from about 19,000 years ago is poor, the optimum growth is at the highest CO_2 level, 700 ppm, about 69% more than the concentration today (415 ppm).

Happer states that plant growth increases proportional to the square root of CO_2 concentration, so going from about 300 ppm to 415 ppm, as we have done over the past century, increases growth rates by around 10%.

Tamblyn's comments:

"A recurring theme in Professor Happer's arguments is expressed in the rather simplistic thought, 'CO_2 is plant food,' which shows an extraordinary lack of knowledge of botany, perhaps understandable since his field is optics. Overwhelmingly, what is missing from his discussion is temperature." (Tamblyn, 2021a, p. 28)

As everyone who took high school biology knows, CO_2 *is* plant food, and one wonders how Tamblyn, an engineer, is so certain of Happer's knowledge of botany? Tamblyn continues:

"Temperature—specifically leaf temperature—is a critical factor in photosynthesis and crop yields. Photosynthesis is temperature-dependent: the productivity of photosynthesis is poor at low temperatures, rising to a peak around 30°C for C3 photosynthesizers, slightly higher for C4 plants. Beyond this peak, photosynthesis efficiency declines markedly, dropping to very low by around 40°C." (Tamblyn, 2021a, p. 30)

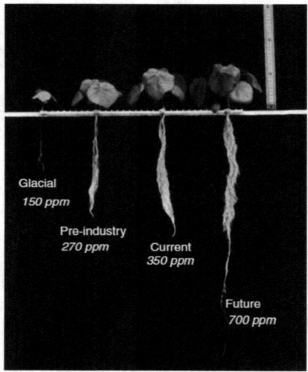

Figure 61. Comparison of plant growth at 14 days from germination under various CO₂ concentrations. Source (Gerhart & Ward, 2010).

Background: CO₂ levels, temperature, and plants

What Tamblyn writes is true and well known, but Happer was careful to say that warming of up to two degrees will be beneficial and Karoly agrees (assuming moderate means about 2°C). The average temperature of Earth's surface today is 14-15°C, well below the optimum temperature of 30°C and much lower than 40°C. Tamblyn also speculates that the nutritional value of plants may be less at higher CO₂ levels, but this is controversial. Research has shown that iron, zinc, and protein levels may decrease about 3% to 10% in some C3 crops at an elevated CO₂ concentration of 550 ppm (Myers et al., 2014). The nutritional value of C4 plants, like corn, are unaffected. How much this small difference will affect human health is unknown. Only when the CO₂

153

concentration reaches 10,000 ppm, which is very unlikely, is it likely nutritional value will decrease significantly (Ward, 2005).

Gavin Foster and colleagues use the discredited IPCC RCP8.5 CO_2-emissions scenario to estimate that if all the fossil fuels on Earth were burned, the CO_2 concentration would rise to 2,000 ppm sometime around 2250AD.[131] They also cite a recent study that speculated CO_2 could rise to 5,000 ppm by 2400AD.[132] Foster, et al.'s CO_2 proxy reconstruction, shown in Figure 62, is much lower than the standard GEOCARB reconstruction shown in Figure 16. GEOCARB tends to follow the higher estimates and Foster's tends to follow the lower estimates. The total range of possible values is very large, for example CO_2 estimates for 200 million years ago range from zero to over 8,000 ppm,[133] so we can only say that CO_2 levels are likely somewhere between those shown in Figures 16 and 62.

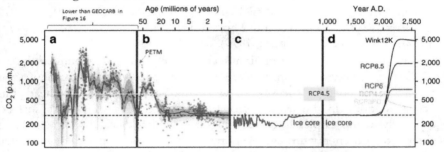

Figure 62. Historical CO_2 from proxies compared to the IPCC AR5 emissions scenarios. This interpretation of CO_2 proxies shows much lower CO_2 prior to one million years ago than the GEOCARB model in Figure 16. After (Foster, Royer, & Lunt, 2017).

The highly speculative projections beyond the year 2100 can be ignored, projections that far in the future are not credible or relevant today. It is well established that the atmospheric CO_2 concentration is very unlikely to exceed 610 ppm in this century.[134] Even so, regardless of the CO_2 reconstruction used, 2,000 ppm is well within the range of CO_2 concentrations seen in the Phanerozoic, so even if it does happen, it is not unusual. We dismiss the extreme emissions projections as too extreme and have highlighted the more reasonable RCP4.5 scenario in yellow, as recommended by Matthew Burgess

---

[131] (Foster, Royer, & Lunt, 2017)

[132] (Lenton, Williamson, Edwards, & al., 2006). Lenton, et al. examine a hypothetical scenario, labeled "Wink12K" in Figure 62, where they model an eventual 12.5°C of warming from 15,000 GtC of emissions. In their speculative model, atmospheric CO_2 reaches 6,000 ppm in 2600AD. Other researchers do not think that level of emissions is possible.

[133] (Foster, Royer, & Lunt, 2017) supplementary materials.

[134] (Wang, Lean, & Sheeley, 2005) show that RCP8.5 is very unlikely to occur.

and colleagues.[135] The RCP4.5 emissions scenario allows more time for the oceans and plants to take up the additional CO_2, and atmospheric concentrations remain lower.

The Paleocene-Eocene Thermal Maximum, which occurred about 56 million years ago, is marked in Figure 62. Atmospheric CO_2 concentration was probably about the same as today, within the margin of error, during the PETM warming event but rose rapidly afterward to levels higher than projected using the RCP4.5 scenario.

Tamblyn's comment about photosynthesis at 40°C, while true, is irrelevant to this discussion and a red herring. The debate is not about what happens at 40°C, but what has happened and will happen at moderate temperature increases. Forty degrees Celsius (104°F) is exceedingly rare for a reason. Over 70% of Earth's surface is water and over open water, it is difficult for the surface temperature to exceed 32°C. This was first shown to be true in a 1979 article by Reginald Newell and Thomas Dopplick in the *Journal of Applied Meteorology* (Newell & Dopplick, 1979).

Energy lost due to evaporation keeps ocean surface temperatures below 32°C nearly everywhere on a clear day in the tropics. The maximum solar energy is absorbed by the surface when the Sun is directly overhead, this occurs at noon in the tropics, along the ITCZ. Most of Earth's surface, especially in the tropics is covered in water, which takes away much of its absorbed solar energy through evaporation. The amount of water evaporated is determined by the vapor pressure of the water, which increases with temperature. On a cloudless day, the balancing temperature is between 27°C and 32°C. Due to the extreme amount of evaporation occurring at 27°C, clouds begin to form rapidly and reflect most incoming solar radiation. Therefore, the tropics warm much less than the poles during warmer times.

All the data we have today—and there is plenty—tells us that as long as there are oceans, the maximum open-ocean surface temperature will be about 32°C. Close to land, in the tropics, and especially in the shallow Persian Gulf, it can exceed that for a short time and even reach 36°C, but over longer periods—especially in the tropical deep ocean—32°C is the maximum. This limits global warming, and it is why global average temperatures have never exceeded 32°C in the past 400 million years (See Figure 11, in Chapter 3).

Richard Willoughby reports that less than 10% of the open-ocean surface water exceeds an annual surface temperature of 30°C. Less than one percent of the ocean surface exceeds 32°C for more than a few days at a time. These warm pools reach their maximum extent in April of each year and are mostly localized in the Indian Ocean and in the tropical Pacific near southeast Asia and around Indonesia (Willoughby, 2021).

---

[135] (Burgess, Ritchie, Shapland, & Pielke Jr., 2020)

Figure 63 shows the average July SST from the NOAA MIMOC[136] dataset. It is a fairly complete global grid and uses data from around 1999 through 2012. The oceans are exceptionally large and even with modern buoys, floats, and ships it takes a while to accumulate enough data to build a worldwide grid at 0.5-degree latitude and longitude resolution. The long and short of it is that 40°C temperatures are virtually impossible over most of Earth's surface, Tamblyn need not be concerned. The maximum temperature in Figure 63 is just over 30°C.

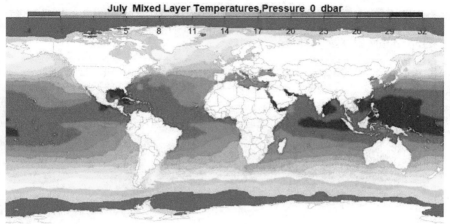

Figure 63. NOAA MIMOC average July SST from approximately 2005 to 2019. The darkest red is the range from 29 degrees to 32 degrees. The nominal depth of these temperatures is 20 cm. Data source: NOAA MIMOC, (Schmidtko, Johnson, & Lyman).

Land-surface temperatures can get much hotter than water surfaces. But water is limited because of evaporation. In Figure 64,[137] gridded weather reanalysis surface temperatures are plotted versus total precipitable water for the globe. Red dots are over land and blue dots are over water. We can see that while land temperatures often exceed 30°C, over water, they do not.

At any point on Earth, there is a water temperature that causes enough evaporation to carry away any excess thermal energy absorbed from the Sun on a cloudless day. North or south of the equator, the balancing temperature is lower, since more incoming radiation is reflected as the angle of incidence increases. The ITCZ forms at the point where the Sun is directly overhead, and it is mostly covered in thunderstorms, these act to cool the ocean surface and carry heat from the ITCZ (Eschenbach, 2016). Because oceans cover over 70% of Earth's surface, and their maximum temperature is limited to about

---

[136] Monthly Isopycnal/Mixed Layer Ocean Climatology
[137] From Willis Eschenbach's blog post "A Chain of Effects," on Wattsupwiththat.com, January 21, 2021.

32°C, the surface temperature of the globe will not be able to exceed that by much.

The ocean surface temperature is often called the "skin temperature." Depending upon the time of day, the wind speed, and whether it is day or night, this "skin layer" is anywhere from a few microns to a few centimeters or so thick (May, 2020d). This is the temperature seen by satellites. During the daytime, with light winds, the skin layer can have a temperature that is two or more degrees warmer than the ocean temperature of the "mixed layer" a few centimeters or so below the surface to the "mixed-layer depth," which varies from 20 meters to 500 meters or so deep. The mixed layer has a uniform temperature and salinity due to surface turbulence.

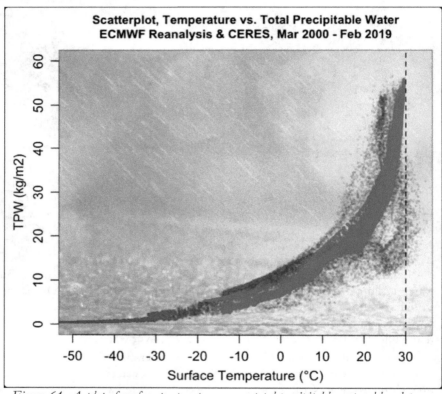

Figure 64. A plot of surface temperature versus total precipitable water, blue dots are over water, red dots over land. Source: Willis Eschenbach, used with permission.

Background: the ocean mixed layer

The average thickness of the mixed layer is 60 to 70 meters, and it has a heat capacity 23 times greater than the entire atmosphere. This means that if the atmospheric temperature were to increase 23 degrees and the heat were transferred to the ocean, which most of it would be, the ocean mixed-layer temperature would increase by one degree. Because the whole mixed layer is in communication with the atmosphere, with a delay of only a few days to a week, it is an effective limit on atmospheric temperatures. See (May, 2020d) for more information on the ocean skin layer and the mixed-layer temperature.

Earth's surface and lower atmosphere are never in thermal equilibrium. Temperatures are always changing as wind and air currents carry heat about and clouds come and go. It is a messy place to measure changes in climate. The ocean mixed layer, due to its large heat capacity and protective skin layer, is more stable and—while in communication with the surface—changes less chaotically.

We will make two more important points about this: One; the overall average surface temperature over the past 550 million years has averaged about 20°C. Throughout that time, it has varied from 11°C to 32°C, thus our current temperature of 14°C to 15°C is unusually cold. Two; the world's surface temperature is effectively buffered by our oceans and has been for 550 million years. As shown in Figure 65, tropical ocean temperatures as measured in the mixed layer and averaged from roughly 1999 to 2019, max out at nearly 29°C—entirely consistent with the theoretical limit of about 30-32°C. Figure 65 plots the *maximum* monthly average temperature for each latitude slice and uses the same data as mapped in Figure 63, however the graph is limited to between 15S to 15N latitude.

Tamblyn's complaint about temperature effects on photosynthesis efficiency is irrelevant. While the oceans exist, it is physically impossible for

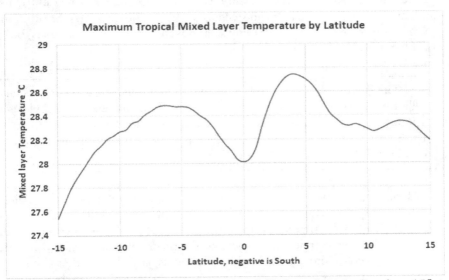

Figure 65. The maximum temperature found in 0.5°-degree latitude slices from 15S to 15N, between 1999 and 2019, in the ocean mixed layer. Data source: NOAA MIMOC (Schmidtko, Johnson, & Lyman).

global average-surface temperatures to exceed 32 degrees Celsius on Earth's surface for any length of time. And the oceans are not going away.

As Happer makes clear, CO_2 is not a pollutant, and more of it will benefit Earth's plant and animal life—including humans. Earth is greener today due to our CO_2 emissions[138] and the result is a three-trillion-US-dollar agricultural benefit to the world (Idso C. , 2013).

Happer writes:

> "Local values of CO_2 can [vary from today's global average of 400 ppm]. For example, exhaled human breath typically consists of 40,000 to 50,000 ppm of CO_2, a fact that should make one wonder about the campaign to demonize CO_2 as a 'pollutant.' Without strong ventilation, CO_2 levels in rooms filled with lots of people commonly reach 2000 ppm with no apparent ill effects. On a calm summer day, CO_2 concentrations in a corn field can drop to 200 ppm or less, because the growing corn sucks so much CO_2 out of the air. The US Navy tries to keep CO_2 levels in submarines below 5,000 ppm to avoid any measurable effect on sailors and NASA sets similar limits for humans in spacecraft." (Happer, 2021b, p. 5)

How does photosynthesis work?

Happer is careful to explain how plants use CO_2 and how additional CO_2 in the atmosphere helps them. Plants combine CO_2, water, a few nutrients, and sunlight to create sugars and cellulose. In the process they breathe out oxygen molecules (O_2). More O_2 in the atmosphere is good for animals, but a problem for plants. As Happer explains:

> "The current low CO_2 levels have exposed a design flaw, made several billion years ago by nature when she first evolved the enzyme, Ribulose-1,5-bisphosphate carboxylase/oxygenase, or 'RuBisCO' for short. RuBisCO is the most abundant protein in the world, and the foundation of all life." (Happer, 2021a, p. 43).

Happer goes on to explain that using energetic molecules, mainly adenosine triphosphate (ATP), produced with the aid of sunlight, "RuBisCO, converts CO_2 to the simple carbohydrate molecule, 3-phosphoglyceraldehyde (3-PGA). The biochemical machinery of the plant subsequently reworks the 3-PGA molecules into sugar, starch, amino acids, and all the other chemicals of life. The letter "C" in the nickname RuBisCO stands for 'carboxylase' in the full name, which reminds us of RuBisCO's design target: CO_2." This

---

[138] (Zhu, Piao, & Myneni, 2016)

concise and informative explanation is quite good for someone with an "extraordinary lack of knowledge of botany." Next Happer explains why excess oxygen is problematic when the atmospheric CO_2 concentration is low.

He explains that "Geological evidence suggests that RuBisCO began to play its key role in photosynthesis some three-billion years ago, when there was a lot of CO_2 and little O_2 in the atmosphere. At current low levels of atmospheric CO_2, plants can use up much of the available CO_2 in full sunlight." As Happer explains, this CO_2 depletion spells trouble for the plant, because if a CO_2 molecule is not found, the plant will grab an oxygen molecule. RuBisCO, charged with chemical energy from ATP, will use the oxygen to make a toxic byproduct, like hydrogen peroxide, instead of useful carbohydrates. This 'photo-oxydation' is a serious problem. At current low CO_2 levels and high O_2 levels, it leads to a reduction of photosynthetic efficiency of about 25 percent in C3 plants, which include major crops: wheat, rice, soybeans, cotton, and many others. Since 3-PGA, the first molecule synthesized from CO_2, has three carbons, such plants are said to have the 'C3' photosynthetic pathway." (Happer, 2021a, p. 43).

Happer then provides us with the illustration in Figure 66 that shows the C3 and C4 processes. In both pathways, CO_2 and H_2O are fused into carbohydrates with the aid of RuBisCO. Harmful O_2 is excluded in C4 plants

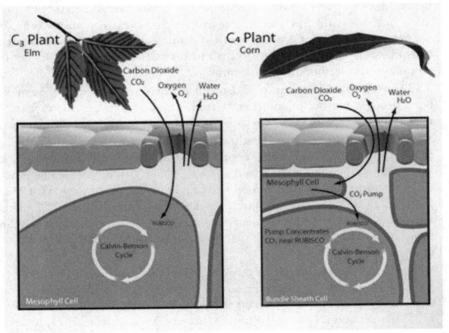

Figure 66. The C3 and C4 processes are illustrated. The opening in the leaves is a stoma. Under low CO_2 conditions leaves have more stomata, which causes them to lose more water to evaporation. Source: (Taub, 2010).

with a pump that concentrates CO_2. Figure 66 also shows a stoma, or hole in the leaf that allows air in and both air and water vapor out.

Most land plants need at least 100 grams of water to produce one gram of carbohydrate. Land plants have finely tuned feedback mechanisms allowing them to grow leaves with more stomata in air poor in CO_2 (like we have today) or with fewer stomata for air richer in CO_2. If CO_2 doubles in the atmosphere, plants reduce the number of stomata in newly grown leaves by about a factor of two. When leaves have fewer stomata, they lose less water to evaporation, and are more drought resistant.

Happer then writes: "A leaf in full sunlight can easily reach a temperature of 30°C, where the concentration of water molecules (H_2O) in the moist interior air of the leaf is about 42,000 ppm, more than 100 times greater than the 400-ppm concentration of CO_2 in fresh air outside the leaf. And CO_2 molecules, being much heavier than H_2O molecules, diffuse more slowly in air. So, depending on the relative humidity of the outside air, as many as 100 H_2O molecules can diffuse out of the leaf for every CO_2 molecule that diffuses in, to be captured by photosynthesis." (Happer, William Happer Interview, 2021a, pp. 45-46).

Karoly has little to say about the importance of CO_2 to life on Earth. Tamblyn complains that some of the areas on Earth that are greener today than in decades past are not arid or warm. He thinks that additional CO_2 only helps in these conditions, forgetting that greener areas deplete CO_2 locally restricting nearby plant growth. Further, plants have moved into some areas that previously were too cold for them to survive, they now grow higher on mountains and farther north and south.

Tamblyn also brings up local drought conditions in his native Australia, but a drought can occur almost anywhere. Planetary greening due to warmer temperatures and more CO_2 is a global event because CO_2 is a well-mixed gas.

Next Tamblyn refers us to the IPCC AR5 Working Group II Report (WGII) Figure SPM.7,[139] which contains a computer model of the change in crop yields from 2010 to 2029 and beyond (IPCC, 2014d, p. 18). The inputs for the model are various emissions scenarios and warming projected for 2029. It shows a slight loss in yield (increases-decreases) from 2010 to 2029. Then after 2029, the losses greatly exceed the gains. But these are not data, they are projections, using model output that does not match the observations shown in Figure 67.

Figure 67 shows crop yields through 2018 from the United Nations Food and Agricultural Organization (U.N. FAO), downloaded from OurWorldinData.org. Yields go up and down over time slightly, but the trend is 97% linear and up about 0.6 tonnes/hectare. Viewed from 2021, it seems

---

[139] The figure is not reproduced here, see Figure 21 in Tamblyn's Detailed Response, page 34 (Tamblyn, 2021a, p. 34).

very unlikely to decrease for the period 2010 through 2029 as predicted by the IPCC.

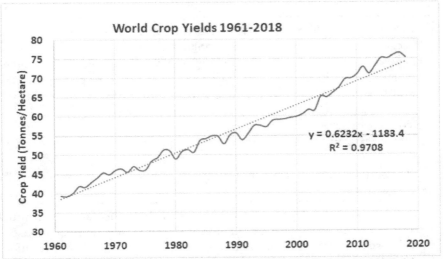

Figure 67. *World crop yields from 1961 to 2018. U.N. FAO data downloaded from OurWorldinData.org.*

In summary, neither Karoly nor Tamblyn offer any evidence that human-caused climate change has adversely affected plant life or will anytime soon. Their evidence, based on models using input from unvalidated CMIP5 climate models, suggests that there *might* be a problem in the far future. Happer and many economists, including Tol, Lomborg, and Nordhaus, have shown that currently the additional CO_2 and warming are benefiting mankind. Idso has calculated a three-trillion-dollar benefit to humans since 1961 and projects that this will grow to $13 trillion by 2050. In the meantime, the world is getting greener, what is not to like?

About the list of scary weather events listed in this chapter that the consensus and alarmists want us to believe are due to human-caused climate change, we need to remember that these are local events and not global. The IPCC admits that no individual weather event can be connected to human-caused warming. Further, there are no data supporting the idea that extreme weather of any kind has increased in recent years.

Extreme weather is always occurring somewhere, and always will. The best plan is for each community or nation to deal with it as it comes. More affluent populations deal with weather extremes better, the best solution is to help the world's population become more affluent, not to decrease fossil fuels which will make them poorer. Next, we discuss the consensus.

Chapter 11: The Consensus

> "[Dr. Tim Ball] watched [his] chosen discipline—climatology—get hijacked and exploited in service of a political agenda, watched people who knew little, or nothing enter the fray and watched scientists become involved for political or funding reasons—willing to corrupt the science, or, at least, ignore what was really going on. The tale is more than a sad story because it set climatology back thirty years and damaged the credibility of science in general." (Ball, 2014, p. 4)

TheBestSchools.org ask Dr. Karoly the following question during his interview (Karoly, 2021a, p. 26):

> "There has been a great deal of discussion about the issue of scientific 'consensus' on the global warming issue. We have already spoken loosely of the 'consensus' as a short-hand way of referring to the official position of the IPCC. However, now we are going to address the broader question of the views of the scientific community at large. One often hears the claim that the consensus among working atmospheric scientists as a whole in favor of the official IPCC position is 'overwhelming;' indeed, one often reads the claim that '97%' of scientists support the mainstream viewpoint. However, some critics question the reality of this consensus on a number of different grounds."

TheBestSchools then references the book *Taken by Storm*, by Christopher Essex and Ross McKitrick.[140] The book begins with this delightfully appropriate quote:

> "In the realm of the seekers after truth there is no human authority. Whoever attempts to play the magistrate, there founders on the laughter of the Gods." Albert Einstein, 1953.

Einstein wrote this after he heard that communist Russia, the USSR, had decided that his theory of relativity was contrary to Dialectical Materialism, the philosophical basis of Marxism. Einstein had long been troubled by the lack of free speech and thought in Russia and wrote the comment above to voice his frustration.

Consensus is a political thing, politicians create a consensus by persuasion, intimidation, and suppressing debate. Science is all about debate, as we see in

---

[140] (Essex & McKitrick, 2003)

this book. In a proper formal debate, the debaters with the best reasoning and facts win. In science a debate may go on for years, centuries even, and is only won with an argument so convincing, supported with data so overwhelming, that *everyone* participating in the debate agrees with it. No consensus is required to convince us that Einstein's theory of relativity is correct, so far, it fits with all pertinent observations. Even so, it may be replaced with something better someday, very few theories survive unmodified forever.

The chilling effect on those who disagree

Essex and McKitrick write that the more the supposed consensus grows, the more it has a chilling effect on scientists who may disagree but prefer to keep their heads down for the sake of their reputations and careers. In this way, consensus is a self-fulfilling prophecy, but one that has little to do with science or the truth.

Karoly responds that:

> "There is an overwhelming consensus of climate scientists who agree that human-caused climate change is happening, and that global warming will continue throughout the current century, with many adverse impacts on human and natural systems." (Karoly, 2021a, pp. 28-29)

Karoly believes that many of the statements by Essex and McKitrick are either wrong or misleading, or both. He writes that scientists cannot control what non-governmental organizations say or what the media report. He believes it is clear that in the U.S. the news media give as much coverage to the small number of skeptical climate scientists as they do to the much larger number of mainstream climate scientists, who discuss the science accurately.

He notes that U.S. politicians are split on party lines in terms of their acceptance of mainstream climate science. Karoly continues, "Any scientist or group of scientists who could reliably and convincingly prove that increasing greenhouse gases in the atmosphere have not caused global warming over the last 100 years and will not cause further global warming would be a strong candidate to win a Nobel Prize in Physics. There is no scientific evidence to support this at present."

The hypothesis being tested is that increasing GHGs are causing, or will cause, dangerous global warming. Yet, Karoly believes the skeptics must prove it is not so. Normally, to disprove a hypothesis, one only needs to find a single example where it does not work. In this case, that has already been done by John Christy and Ross McKitrick. Their 2018 paper provides a statistically valid refutation of all IPCC climate models.[141] It shows the climate models' tropical hot spot is warming faster than observations, as explained in Chapter

---

[141] (McKitrick & Christy, 2018)

5. Although the 2018 paper was published after the debate, Christy's evidence that the models were invalidated by tropical troposphere measurements was widely available in 2013.

Apparently, this is not sufficient in this case, and Karoly does not explain why. Karoly's statement can be reversed, since any scientist who could convincingly prove that humans are causing climate to change will be a strong candidate to win a Nobel Prize. There is obviously no convincing proof of this assertion today, as this debate demonstrates.

Karoly explains that the "consensus is based on a vast body of scientific evidence and many thousands of peer-reviewed scientific publications that have been assessed in the [IPCC] reports. It is the consensus of the evidence and the peer reviewed publications that is important, which leads to the consensus of climate scientists." (Karoly, 2021b, p. 3). This reminds us of the Queen's statement to the tedious Polonius in William Shakespeare's *Hamlet*. "More matter with less art." What she means is more substance and less rhetoric. The IPCC has produced thousands of pages of rhetoric from thousands of papers, yet the uncertainty in the calculated effect of human GHG emission on our climate is the same three degrees it was in 1979. Burying us in paper does not change that basic fact or narrow the large uncertainty.

There are many hundreds, if not thousands, of peer-reviewed publications challenging the consensus view. He does not explain why these do not matter. In the scientific community, it only takes one paper to challenge a hypothesis like human-caused climate change. The number on each side does not matter, only the replicability of the results.

TheBestSchools mentions that France's Mont Blanc glacier and Alaska's Hubbard Glacier have advanced, which Karoly disputes. Karoly is quite correct that these glaciers have retreated in the past two-hundred years or so, but they advanced for several thousand years before that. The Mont Blanc glacier reached its maximum Little Ice Age extent in 1818 (Nussbaumer & Zumbuhl, 2012). Other recent maxima were reached in 1921, 1941, and 1983. The glacier has been retreating since 1983.

TheBestSchools asked Happer the same question, "Is there a consensus on global warming? Is it overwhelming? His response:

> "Essex and McKitrick are on target in their book, *Taken by Storm*. It is striking that many skeptics, like me, are retired. Aside from character assassination, there is not much the attack dogs of the climate consensus can do to us, at least so far. But young academics know very well that they will risk their careers by expressing any doubt about the party line on global warming. (Happer, 2021a, p. 47)

Happer then provides a portion of Michael Crichton's famous quote on the consensus from his California Institute of Technology speech in 2003. We also quoted a small portion in Chapter 5, below is more of what he said:

"I want to pause here and talk about this notion of consensus, and the rise of what has been called consensus science. I regard consensus science as an extremely pernicious development that ought to be stopped cold in its tracks. Historically, the claim of consensus has been the first refuge of scoundrels; it is a way to avoid debate by claiming that the matter is already settled. Whenever you hear the consensus of scientists agrees on something or other, reach for your wallet, because you're being had.

"Let's be clear: the work of science has nothing whatever to do with consensus. Consensus is the business of politics. Science, on the contrary, requires only one investigator who happens to be right, which means that he or she has results that are verifiable by reference to the real world. In science consensus is irrelevant. What is relevant is reproducible results. The greatest scientists in history are great precisely because they broke with the consensus.

"There is no such thing as consensus science. If it's consensus, it isn't science. If it's science, it isn't consensus. Period.

"... Finally, I would remind you to notice where the claim of consensus is invoked. Consensus is invoked only in situations where the science is not solid enough. Nobody says the consensus of scientists agrees that $E = mc^2$. Nobody says the consensus is that the sun is 93 million miles away. It would never occur to anyone to speak that way." Michael Crichton, 2003.

Happer then further discusses how extreme "consensus science" can get:

"Consensus supporters don't like to admit it, but the situation is getting perilously close to Lysenkoism. Lysenko ... gained complete control over biology in the USSR, with the full backing of the Politburo and the personal support of both Stalin and Khrushchev.

"... Scientists who expressed any doubt about Lysenko's dogmas were lucky if they were only fired from their jobs. Many were sent to concentration camps in Siberia, and some were sentenced to death." (Happer, 2021a, p. 47)

If one scientist describes his work and another, independent scientist does the same work with the same results, that's science. Ten million scientists' opinions are nothing in comparison.

Tamblyn does not directly discuss the consensus but does mention how many national scientific bodies and governments support the idea that humans are causing dangerous climate change (Tamblyn, 2021b, p. 24). That governments would support the so-called consensus is expected, they are political bodies. That national scientific organizations do the same, rather than supporting scientific debate, is a tragedy.

166

Chapter 12: Is Government Intervention Necessary?

"The prospect of domination of the nation's scholars by Federal employment, project allocation, and the power of money is ever present and is gravely to be regarded." President Dwight David Eisenhower's Farewell Address January 17, 1961, (Eisenhower, 1961).

President Eisenhower foresaw the climate-change problems decades before they happened. Fear is the most powerful motivator a politician has, and by gathering thousands of scientists with promises of endlessly renewable research grants, a politician can entice them to create all sorts of disaster scenarios to keep the public alarmed. In the words of H. L. Mencken:

"The whole aim of practical politics is to keep the populace alarmed (and hence clamorous to be led to safety) by menacing it with a series of hobgoblins, all of them imaginary." (Mencken, 1918).

In today's world of computers, the lack of evidence that climate change is caused by humans or dangerous is not a problem. Evidence can be conjured up, as if by magic, with computer models. As Mark Twain wrote:

"In the space of one hundred and seventy-six years the Lower Mississippi has shortened itself two hundred and forty-two miles. That is an average of a trifle over one mile and a third per year. Therefore, any calm person, who is not blind or idiotic, can see that in the 'Old Oolitic Silurian Period,' just a million years ago next November, the Lower Mississippi River was upwards of one million three hundred thousand miles long, and stuck out over the Gulf of Mexico like a fishing-rod. And by the same token any person can see that seven hundred and forty-two years from now the Lower Mississippi will be only a mile and three-quarters long, and Cairo and New Orleans will have joined their streets together and be plodding comfortably along under a single mayor and a mutual board of aldermen. There is something fascinating about science. One gets such wholesale returns of conjecture out of such a trifling investment of fact." Mark Twain in *Life on the Mississippi,* Chapter 17 (Twain, 1883).

Mark Twain was also prescient. Little did he know, that a little over 100 years later, the IPCC would "conjecture" in much the same way. They just used computer graphics rather than clever words.

TheBestSchools asked Happer if he thought government intervention was necessary to combat anthropogenic global warming, Happer's response:

"This is nonsense. Government actions to combat the non-existent problem have blighted the landscape with windmills and solar farms. They have driven up the price of electricity, which has disproportionately harmed the poorest segments of society. Government actions have corrupted science, which has been flooded by money to produce politically correct results. It is time for governments to finally admit the truth about global warming. Warming is not the problem. Government action is the problem." (Happer, 2021a, p. 20)

Should the IPCC correct their CO$_2$ sensitivity estimate?

Happer believes that if the IPCC corrected their exaggerated ECS to CO$_2$, they would find that no government intervention is needed. He cites a 2013 *Nature Climate Change* paper by John Fyfe, Nathan Gillet and Francis Zwiers entitled "Overestimated global warming over the past 20 years" documenting that climate models have significantly overestimated warming from 1993 to 2012.[142] By significantly, they mean a factor of two, and note that only a few simulations provide warming trends within the range of observational uncertainty. With the period narrowed from 1998 to 2012 (the Pause), the overestimation becomes worse by a factor of four.

Because the models overestimate warming, they also overestimate the impact of fossil-fuel CO$_2$ emissions, because natural influences on climate are assumed to net to zero over the recent (<60 years) past. The following is Happer's answer to Tamblyn's laundry list of potential man-made climate hazards (Tamblyn, 2021a, pp. 18-37).

"Is concerted governmental action at the global level desirable? No. More CO$_2$ will be good for the world, not bad. Concerted government action may take place anyway, as has so often happened in the sad history of human folly." (Happer, 2021a, p. 49)

Happer's point is valid. There are no observations or measurements suggesting human CO$_2$ emissions are causing warming, only models. Even if CO$_2$ is causing some warming, economic models show it is not a problem.

Tamblyn does not believe the excess predicted warming by CMIP6 models (Figure 28, Chapter 5), and in similar figures in Happer's statement, are over long enough periods to draw any conclusions. The models also overestimate

---

[142] (Fyfe, Gillett, & Zwiers, 2013)

climate sensitivity to rising CO_2. Tamblyn justifies this by providing more estimates. They are shown in Figure 68.

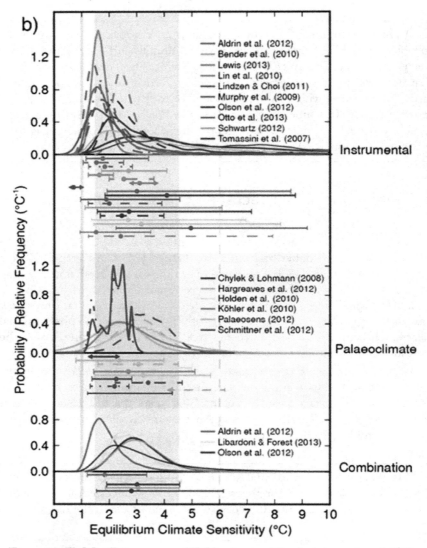

Figure 68. IPCC AR5 summary of ECS estimates. The shaded area from 1.5°C to 4.5°C represents the IPCC's AR5 official range. Source AR5 (IPCC, 2013, p. 925).

Tamblyn acknowledges that climate sensitivity is unknown but suggests the estimates "cluster around 3." That is not very precise. The IPCC's official range of ECS in AR5 was 3.0°C ±1.5°C per doubling of CO_2, the same range given in the Charney Report in 1979.[143] In AR6, their very likely range is from

---

[143] (Charney, et al., 1979)

two to five degrees,[144] and they admit the CMIP6 model ensemble has a broader range of estimates of climate sensitivity than the CMIP5 model ensemble in 2013.[145]

The AR6 best estimate of ECS is three degrees,[146] precisely where it was in the Charney Report. After 40 years of work and billions of dollars spent, the ECS—the key number that tells us what the warming from additional CO_2 will be—has not changed or become more accurate. AR6 makes the point that while their *very likely* range of values is two to five degrees, their *likely* range is 2.5 degrees to four degrees and this is a smaller range that the three-degree uncertainty given in AR5, but this is just semantics.

Figure 68 is like one found in Tamblyn's detailed response (his Figure 8, page 16, Tamblyn's detailed response), but ours is the version used in AR5.[147] The main takeaway is that the IPCC has no clue how much CO_2 affects the climate and atmospheric warming. The ECS estimates shown, range from less than one to over nine degrees Celsius!

Tamblyn uses Figure 68 hoping to illustrate that Happer's opinion stating ECS is below the IPCC estimate of "about 3" is wrong. How would Tamblyn, or anyone, know? Many of the estimates in Figure 68 are less than two degrees and the main body of estimates covers over four degrees. The figure only shows how little we know about the impact CO_2 has on climate.

The IPCC AR6 official "best estimate of equilibrium climate sensitivity (ECS) of 3°C, with a very likely range of 2°C to 5°C" is an exceptionally large uncertainty for such an important number. A value of two is benign, meaning we have nothing to worry about. A value of five might be dangerous for some, hard to say. These numbers are not accurate enough to make critical decisions. If we are to disrupt lives, throw millions of people out of work, raise electricity costs, and spend trillions of dollars replacing fossil fuels, shouldn't we be more precise about the dangers? This is not an incentive for government intervention.

Climate Sensitivity

Because the estimates of the effect of CO_2 (ECS) on climate range from a beneficial ($<2°C/2xCO_2$) to a possibly dangerous ($>4°C/2xCO_2$), the necessity of government intervention depends directly on the estimate of ECS. The debate is not about whether human CO_2 emissions affect climate change, it is about how much. Professor Happer believes climate sensitivity is low and writes the following in his final reply:

---

[144] (IPCC, 2021, p. TS-39)
[145] (IPCC, 2021, p. TS-18)
[146] (IPCC, 2021, p. TS-57)
[147] Figure 10.20b, page 925, AR5

> "It is immoral to deprive most of mankind of the benefits of affordable, reliable energy from fossil fuels on the basis of computer models that do not work." (Happer, 2021d)

Here, Happer is stating, accurately, that the climate models have not been validated with a proper test of their accuracy.

TheBestSchools posed the following question to Karoly in his interview:

> "Let us begin by pointing out that it is logically possible to acknowledge that global warming is a reality, is manmade, and is a very bad thing—and yet to feel that the sort of concerted global governmental intervention that is being proposed by the IPCC and the UN Conferences on Climate Change is a cure worse than the disease." (Karoly, 2021a, pp. 29-30)

Karoly begins his response by pointing out that all governments of the world agreed to the Paris agreement in 2015 and to the United Nations Framework Conventions on Climate Change (UNFCCC) in 1992. The UNFCCC objective was the "stabilization of greenhouse-gas concentrations in the atmosphere at a level that would avoid dangerous anthropogenic interference with the climate system." In other words, government officials agreed that governments should be given more power, so they can, supposedly, control the world's climate. When has a government official turned down additional powers?

Karoly continues:

> "It is also important to clear up the misunderstanding that the IPCC makes recommendations and provides policy advice to governments. This is incorrect. The IPCC undertakes comprehensive assessment of the peer-reviewed literature about climate change, including climate change science, projections of future climate change, impacts of climate change, vulnerability and adaptation to those impacts, and approaches to reducing climate change due to human activities. ... [The] Summaries for Policymakers ... are required to be policy-neutral. ...
>
> "Any decision on how to respond to climate change is an ethical and moral decision for society, and not just an economic decision." (Karoly, 2021a, p. 31)

Both international agreements have provisions that call for wealthier countries to provide money and resources to developing countries and neither agreement commits any country to do anything. What's not to like?

Karoly's statement is true, on its face. But, nonetheless, the IPCC has a long history of slanting or changing its *scientific* conclusions to match its own *political* objectives as well as those of individual governments. As discussed in Chapter 1, their corruption of the scientific process goes all the way back to the second IPCC report in 1995 when the conclusions of Chapter 8—the chapter attributing global warming to humans—were changed at the insistence of politicians and without the permission of the scientists who wrote and agreed to it (May, 2020c, pp. 230-238).

A government—possibly the U.S. government—insisted that Chapter 8 explicitly say that humans caused climate change. The leader of the SAR effort, John Houghton, insisted that IPCC procedures were followed, and that governments (meaning politicians), as opposed to scientists, had the final say (Houghton, 1996). The IPCC reports are scientific, or they are political; they cannot be both. The scientists already agreed that they could not be sure humans controlled the climate, because they did not know the extent of natural variability. The politicians insisted this be changed, and it was. This incident, more than any other in the history of the IPCC, destroyed their reputation as an independent and unbiased reporter of the state of climate science.

In Karoly's view, evidence exists confirming global warming is currently affecting human health and sea level rise is leading to coastal flooding. He also believes global warming is affecting natural ecosystems around the world—especially coral reefs in the tropics—and will lead to the extinction of growing numbers of species. He believes the destruction he foresees is immeasurable (Karoly, 2021a, p. 31).

Karoly sees all of this as a moral and logistical problem. He clearly thinks global warming will affect the tropics more than the higher latitudes. However, the tropics have not warmed as quickly as the Northern Hemisphere (Figures 8 and 9, Chapter 2). As Figure 9 shows, the tropics have warmed at 85% of the Northern Hemisphere rate since 1979. Thus, global warming is not occurring uniformly across the planet. This becomes difficult to explain when CO_2 is supposedly a well-mixed gas that disperses quickly around the globe and internal climate variability supposedly nets to zero.

Karoly appears to recognize this in his major statement when he writes:

> "It is important to recognize again that the natural variability of surface temperatures is much greater at regional and local scales than at the global scale. Hence, if we wish to identify the possible causes of global warming, it is important to consider variability of global or hemispheric average temperatures, and not regional temperature variability." (Karoly, 2021b, p. 11)

Only focus on global and hemispheric variability when looking for the cause of climate change? Some naturally caused changes will be seen at the hemispheric level over a century-long time scale, but few, if any, will be seen at the global level. If we limit ourselves to looking at global average temperature, we dictate the answer, since natural forcings differ by latitude. Only CO_2, among the current options, is global.

Background: Nordhaus and past government interventions

William Nordhaus, the Nobel Prize winning Yale economist, has analyzed the economic consequences of combating climate change through a reduction of fossil fuels more than just about anybody. Much of the following summary is from his Nobel Prize acceptance speech in Stockholm (Nordhaus W., 2018).

The first government action was to form the UNFCCC in 1994. The stated goal was to "achieve ... stabilization of greenhouse gas concentrations in the atmosphere at a level that would prevent dangerous anthropogenic interference with the climate system." Thus, built into the goal was the unproven idea that we humans can somehow dangerously alter the climate with our greenhouse-gas emissions.

The first solid step toward their goal was the Kyoto Protocol of 1997. High-income countries agreed to reduce their emissions to five percent below 1990 levels by 2012. The protocol created an international cap-and-trade emissions system. It was an ambitious plan, but few joined, and no one complied with the rules. Most countries eventually backed out, and one of the first was the United States. Basically, it was a "club no country cared to join" (Nordhaus W. , 2018). Kyoto died an early death in 2012, mourned by few.

Then came the Paris agreement in 2015. This agreement set a target of two degrees Celsius above pre-industrial levels. All countries agreed to make their "best effort." There were no penalties for withdrawing from the agreement or failing to achieve the goals. The agreement drew a lot of attention but accomplished nothing.

Nordhaus thinks the best measure of the effectiveness of these three agreements is what he calls "carbon intensity," which is the ratio of CO_2 emissions to output. In 2010, the U.S. emitted 5.7 billion tons of CO_2, with a real GDP of \$14.8 trillion yielding a carbon intensity of 0.386 tons/\$1,000 of GDP. By 2015, it fell to 0.328 tons/\$1,000 of GDP, a decline of three percent per year.

If the past agreements were effective, then the trend of carbon intensity would have declined after each was signed. Yet, the record shows no change following any of them. In fact, most of the global decarbonization improvement since 1980 resulted from the modernization of China. With China excluded, the rate of decarbonization of the global economy from 1980 to 2017 was a steady 1.8% to 2.0% per year. Global productivity, or GDP per worker increases at this rate, so a 2% increase is essentially zero. So, except for China, we are in the same place we were in 1980. Nordhaus's Figure 6 (our Figure 69) illustrates this. The agreements were added to his figure. Because this is a plot of economic ratios, the Y axis must be plotted logarithmically.[148] Figure 69 is the global economy, it includes China.

In Figure 69, the overall improvement in carbon intensity from 1960 to 2017 is 1.6%/year. From 2010 to 2017, the improvement is 2.3%/year. China's recent rapid 2010–2017 modernization led its carbon intensity to drop at a rate of 5%/year. If China is excluded from 2010 to 2017, the global improvement falls to 2.0%/year. As a result, Nordhaus concludes the climate agreements have been ineffective.

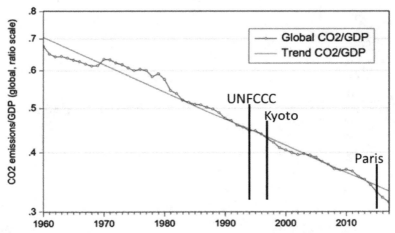

Figure 69. Global CO_2/GDP versus the linear trend. The global CO_2 efficiency, including China, is on a nearly linear trend, with no significant improvement in decades. After: (Nordhaus W., 2018).

Global GDP is increasing about three percent per year. Most of this improvement is due to productivity improvements, especially in China where industrialization is rapid. Thus, CO_2 emissions world-wide are increasing, just not as fast as productivity. CO_2-emissions growth has shown no sign of decelerating and is increasing steadily as shown in Figure 70. Emissions are increasing linearly, have nearly quadrupled since 1960, and the UNFCCC, Kyoto and Paris agreements are not visibly making any difference.

Nordhaus emphasizes CO_2 efficiency, but is CO_2 truly dangerous? The absolute numbers (shown in Figure 70) matter and they are staying on track. Nordhaus admits the past agreements have been failures and have not influenced CO_2 emissions. However, he is optimistic that an agreement could be forged that would make a difference. Nordhaus emphasizes that any agreement must punish "free riders," those countries that do nothing but benefit from the sacrifices made by the others.

---

[148] (Rothman, Wise, & Hatch, 2011)

He recommends an agreement he calls a "club." He cites some important international clubs including NATO, the European Union, and other multinational trade agreements. But we would argue that these clubs have been successful because they were built to fight real and clear threats. The Soviet Union was a real threat, making NATO a necessity, and trade agreements were forged to fight counter-productive tariff wars that were leading to inflation and lower standards of living.

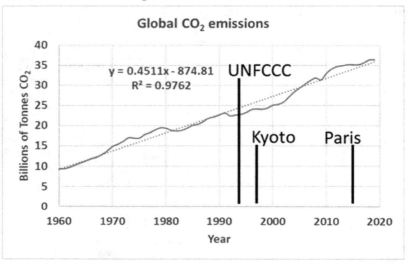

Figure 70. Global CO₂ emissions from OurWorldinData.org. The trend is very linear and has an R² of 0.98.

There are no observable or measurable data that climate change is dangerous, only models and conjecture. Until there is real evidence, the agreements will probably continue to be meaningless and ineffective. Models are not evidence, just ideas programmed into a computer. Nordhaus's analysis is rigorous, but by assuming an implausible CO_2 emissions scenario (RCP8.5), his costs of inaction are too high. If he were to use a more plausible scenario, say RCP4.5, he would find that the costs of inaction are low and hardly worth messing with. We will look at this in more detail in the next chapter.

Nordhaus is convinced that global warming is a serious threat, regardless of the lack of observations or measurements of the effect. He accepts the IPCC conclusions. It is an extremely common opinion.

In Nordhaus's view, the problem with the Kyoto and Paris agreements is they focused on CO_2 emissions reductions. He thinks that the focus should be on a globally agreed upon carbon dioxide tax. He concludes that a price of $25/ton for carbon is best and should be applied worldwide. This would eliminate free riders, but would also reduce economic growth, especially in poorer countries. Everyone must be convinced that global warming is a true threat, with real dangers, before this sort of solution can work. Karoly,

Nordhaus, and Tamblyn are convinced, Happer and this author are not. We examine more of Nordhaus's Nobel Prize speech in the next chapter.

Chapter 13: Should we do anything about Global Warming?

"Up to ten years ago, I [repeated] what the IPCC tells us, [without checking it]. Then at some point I started to check the allegations. The result: it started with doubts and then ended in horror, in the realization that a great deal of what the IPCC and the media are saying about climate change is incorrect and is not [supported] by scientific facts and measurements. I am ashamed of what I used to 'recount' in my own lectures as a natural scientist. ...

"From a scientific point of view, it is almost absurd to [think] some CO_2-adjusting screws [produce] a nice, pleasant, stable climate." Physicist and meteorologist Dr. Klaus-Eckart Puls, translated from the original German (Hahne-Waldscheck, 2012).

Karoly thinks that "limiting global warming to any level requires stabilizing greenhouse-gas concentrations in the atmosphere." Due to the unmeasured but computed climate sensitivity, Karoly believes "rapid, substantial and sustained reductions in greenhouse-gas emissions from human activity" are required. He continues:

"The net emissions (sources minus sinks) of greenhouse gases into the atmosphere from human activity need to fall from present levels to near zero as quickly as possible." (Karoly, 2021b, p. 24)

CO_2 and climate

Karoly provides our Figure 71 below as an illustration of the impact of human CO_2 emissions on climate. The plot is model-based and assumes a range of CO_2 emissions scenarios that cover the shaded region. The warming shown is for the year 2100. An unstated assumption used to make the graph is an implied climate sensitivity of approximately 2.2°C/2xCO_2. The year 2100 CO_2 concentrations on the plot imply a TCR as opposed to an ECS. The difference being that ECS is the temperature-rise reached after the oceans have equilibrated with a new surface temperature, which takes several hundred years to achieve, whereas TCR is the change in temperature due to CO_2 over a shorter term—in this case 100 years. The CS is not given in the text. The AR5 model-mean (page 818) ECS is 3.2 and the mean TCR is 1.8. If the CS is lower than assumed by Karoly, or the IPCC, the slope of the ellipses and the shaded region will be lower than illustrated and the resulting temperatures in 2100 will be lower at each level of CO_2 emissions.

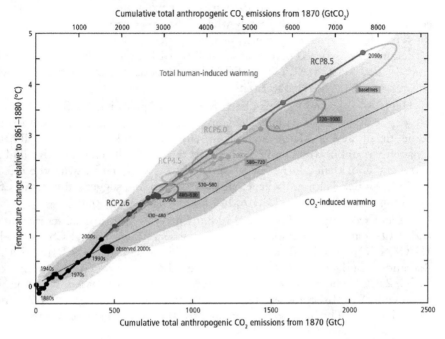

Figure 71. CO₂ *emissions versus temperature. From IPCC AR5, Synthesis Report Figure 2.3 (IPCC core writing team 2014). The ellipses show the temperature change of various IPCC emission scenarios, the numeric labels are ranges of* CO₂ *concentration in 2100. GtCO₂ = gigatons CO₂.*

Quoting the AR5 report "Summary for Policymakers":

> "Equilibrium climate sensitivity is likely in the range 1.5°C to 4.5°C (*high confidence*), extremely unlikely less than 1°C (high confidence), and very unlikely greater than 6°C (medium confidence)." (IPCC, 2013, p. 16)

The above quote has a footnote that reads as follows:

> "No best estimate for equilibrium climate sensitivity can now be given because of a lack of agreement on values across assessed lines of evidence and studies."

The footnote admits that the model results and observation-based estimates of ECS do not agree with one another. We, as humans, are more interested in the TCR, a smaller value representing the temperature rise in degrees Celsius over a period in which CO_2 increases steadily, generally at 1% per year, when the concentration has doubled. TCR is the rapid (70–100 years or so) temperature response at the surface, at this point, there is no wait for equilibrium. TCR is approximately what Happer is referring to when he uses the term climate sensitivity, which he abbreviates as "CS," or sometimes "S." Generally, it is assumed that TCR is 50–80% of ECS according to Lewis and Curry[149] and IPCC's AR5. The difference depends mainly on the assumed ocean response to surface warming.

Happer estimates a feedback-free, pre-equilibrium CS of about $1°C/2xCO_2$ and explains his calculation (which we cover in Chapter 16) in some detail in his Major Statement. How feedbacks affect the climate sensitivity is unknown. The IPCC believes the net feedback is positive and is about 2.1°C of their central ECS estimate of 3.2°C—that is, twice the effect of CO_2 alone. Happer believes that observations suggest the total CS, including all feedbacks, is closer to his theoretical value of one.

Recent work by Nic Lewis and Judith Curry, using historical temperature and CO_2 data,[150] conclude, based on observations, that ECS is likely $1.5°C/2xCO_2$ and TCR is $1.2°C/2xCO_2$. Lindzen, using ERBE (Earth Radiation Budget Experiment) satellite data, computed an ECS of $0.5°C/2xCO_2$ and believes the net feedback on CO_2-based warming is negative.[151] So, observations suggest feedback is between a negative 30% and a positive 20%. The high IPCC estimate probably causes them to overestimate future global warming.

The observation-based estimates of CO_2-caused warming by Lewis, Curry, Lindzen, and Choi are benign and, if true, will be beneficial to humans, plants, the economy, and animal life. Even the IPCC estimates are not dangerous to human health. Although, if the IPCC is correct, four or five degrees of warming might have economic repercussions. So next we look at estimates of the economic impact.

Background: CO_2 costs and benefits

Richard Tol reviewed numerous papers on the economic impact of global warming and found that warming to 1.1°C above pre-industrial times is beneficial and the societal benefits do not become negative until warming exceeds about two degrees Celsius (Tol R., 2015c). If Happer and Lindzen are correct, and we reach that level of warming before 2100, it will not be CO_2 that did it.

---

[149] (Lewis and Curry 2015)

[150] (Lewis and Curry 2018)

[151] (Lindzen & Choi, 2009), in a later paper (Lindzen & Choi, 2011), they increased their ECS estimate to 0.7°C, but the uncertainty is 0.5 to 1.3°C, so we use 0.5°C as a lower bound in this discussion.

Further, economic growth does not stop under any scenario. Bjorn Lomborg reports the IPCC has estimated that 60 years of unmitigated global warming would cost the planet between 0.2% and 2% of global GDP—hardly the end of the world (Lomborg B. , 2018). According to the International Monetary Fund's World Economic Outlook, the current global growth of GDP is about three percent per year. 2019 global GDP was about $88 trillion (statista.com). So, three percent real growth per year with 2019 GDP at $88 trillion yields a GDP of $519 trillion in 2080. Even if global warming cost the IPCC projected maximum of two percent, global GDP will still be $509 trillion dollars in 60 years, more than five times what we produce today. The cost difference is insignificant.

Lomborg points out that Nordhaus calculated the economically optimal outcome is warming of 3.5°C and that reducing temperature rises by more through CO_2 mitigation efforts would result in higher costs than benefits, potentially costing the world $50 trillion in the short term. The problem with the IPCC calculations, is they have not properly factored in CO_2 mitigation costs. The suggested solutions are worse than the warming, even if we assume—without proper observational evidence—that CO_2 is causing most of the warming.

As mentioned in the previous chapter, Nordhaus presented his Nobel Prize talk at the Aula Magna, Stockholm University, December 8, 2018, (Nordhaus W. , 2018). Nordhaus received his Nobel Prize for work on climate change economics (The Nobel Prize, 2018). In the talk he assumed the unrealistic and implausible RCP8.5 business-as-usual scenario (Hausfather & Peters, 2020). Interesting blog posts about RCP8.5 were penned by Curry and Kummer (Curry J., 2018b). Jianliang Wang and colleagues have shown that there is not enough coal in the world to allow the RCP8.5 scenario to occur.[152]

But Nordhaus used it to do his analysis anyway. He did his best to project the global economic benefits and costs of both the IPCC projected global warming and various CO_2 (fossil fuel) mitigation ideas, something the IPCC has yet to do. Figure 72 plots the projected temperature changes for all of Nordhaus's economic scenarios.

The economically optimum scenario (orange dotted line with triangles) reaches four degrees of warming before leveling out. It is the scenario with the lowest cost to society. The increase in temperature is higher than the IPCC recommends we plan for, and considerably higher than observations suggest we will reach.[153] It is also higher than Happer projects. Four degrees of warming is certainly not physically dangerous to humans, animals, or plants. Life thrives in areas with more extreme temperatures than that, so our economy is the main concern, and Nordhaus is saying four degrees of warming is optimal from an economic perspective.

---

[152] (Wang, Feng, Tang, Bentley, & Höök, 2017)
[153] (Lewis & Curry, 2018) (Lindzen & Choi, 2011)

Nordhaus arrives at his conclusions by analyzing the "social cost of carbon" dioxide emissions. If a carbon tax is applied to fossil fuels, then the cost of the fuels goes up causing damage to the economy. Fossil fuels are used to make or distribute nearly everything we use in our daily lives; thus, implementing the tax increases the cost of everything, which reduces consumption and lowers our standard of living. But, if the IPCC's worst-case scenario is true, and their analysis of the cost of warming is accurate, there will be benefits to raising the cost of fossil fuels because we will use less of them, emit less CO_2, and avoid costs due to global warming. Nordhaus plots the damages and benefits of various CO_2 mitigation taxes and compares them to IPCC estimated costs of warming in Figure 73.

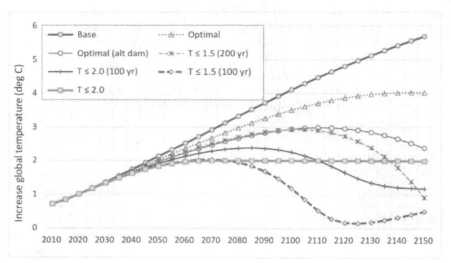

Figure 72. William Nordhaus's temperature scenarios for his analysis of the costs and benefits of combating global warming. The economically optimal path is identified with orange triangles, and it hits a warming of 4°C before leveling off. The costs and benefits for each temperature scenario are plotted in Figure 73. The optimal "alt dam" line is like the orange line, but with an alternative damage (aka warming) function.

The IPCC would like us to limit warming to 1.5°C over the pre-industrial era, roughly pre-1900, when CO_2 was about 280 to 290 ppm. But, as Nordhaus's analysis shows (see the right side of Figure 73), this would be an economic disaster and cost over $50 trillion. It would likely impoverish nearly everyone. Even limiting us to less than two degrees in the next 100 years, would lead to economic disaster. Compare this with the IPCC estimate of a trivial 0.2–2% reduction in GDP, in 60 years, if we do nothing.

Considering that RCP8.5 is very unlikely, Nordhaus's analysis suggests that doing nothing about global warming is the best solution. Individuals and communities can adapt to any local changes due to warming, as we always have.

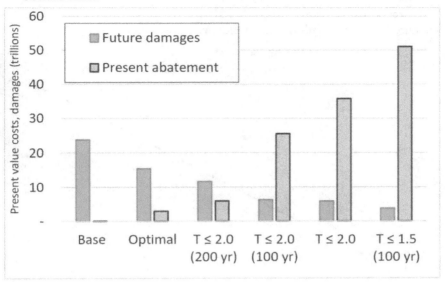

Figure 73. The cost of future economic damage due to warming in today's dollars in green, and the present cost of CO₂ mitigation or abatement in red (Nordhaus W. , 2018). These are the same temperature scenarios plotted in Figure 72, except for the alternative damage ("alt dam") function.

The IPCC, and most economists, believe warming of two degrees from the pre-industrial era (or about 1.1–1.2°C above today) is unlikely to cause problems and may be beneficial long term, but how long does it take for warming of two or four degrees to occur if we continue adding CO_2 at the current rate? We can compute it with Figure 74.

With the various proposed values of climate sensitivity to CO_2 in mind, Happer provides us with an interesting and informative graph in his Major Statement. Figure 74 is a modified version.

We see from the gray line in the Figure, that using the IPCC central estimate of transient climate response of 1.8°C/2xCO₂,[154] we reach an increase of two degrees from the pre-industrial era, in a little over 80 years from today—roughly 2100. The warming curves start at 290 ppm, the pre-industrial CO_2 concentration. The black line projects the current (2020) rate of increase in CO_2 concentration of 2.5 ppm/year. We see the concentration needed where the gray line crosses two degrees on the left scale, then see where that concentration falls on the black line and read the number of years on the right-hand scale.

Using the observation-based TCR estimate of 1.2°C/2xCO₂ (Lewis & Curry, 2018), the orange line, we do not reach two degrees until a little over

---

[154] (IPCC, 2013, p. 818)

200 years from today—roughly 2220. This is where the concentration required, 900 ppm, falls on the black line.

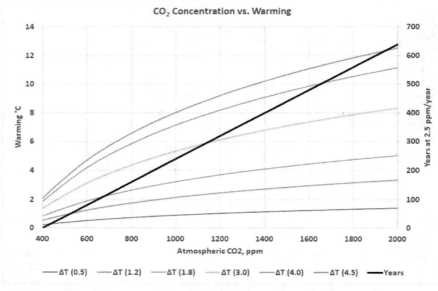

Figure 74. A graphical description of the effect of climate sensitivity and the current rate of CO₂ growth (2.5 ppm/year) on warming. The black line is years from 2020, at a rate of 2.5 ppm per year (right-hand scale). The curves are warming from the pre-industrial CO₂ concentration of 290 ppm using the left scale. Each line represents a different climate sensitivity. After Happer's Figure 12 in his Major Statement (Happer, 2021b, p. 29).

As discussed previously, ECS is significantly delayed, by hundreds of years because of the time it takes for ocean temperature to equilibrate with the atmospheric temperature. So, the more appropriate sensitivity to use is TCR, which is still delayed, but less than 70 years or so. The other lines cover other emissions scenarios. The darker blue line is roughly the Lindzen and Choi lower warming scenario of 0.5°C/2xCO₂, the yellow, light blue, and green lines show progressively more extreme and unlikely warming scenarios from the IPCC reports.

No one has ever measured the warming, or feedbacks, due to CO₂ emissions in nature. The IPCC AR5 central estimate of TCR is representative of those derived from models and the Lewis and Curry estimate is representative of those derived from observations. Using these estimates and the current rate of CO₂ increase, any potentially negative effects of global warming are, at least, 80 to 200 years in the future, possibly much longer.

What will the world be like then? Eighty years ago, we had no integrated circuits or jet aircraft. Two-hundred years ago, trains had just been invented and there were no automobiles or electric light bulbs. Eighty years from now we will be much wealthier and have technology we can't even dream of today.

Dr. Nordhaus found an optimum economic scenario that called for reaching four degrees of warming from pre-industrial times. This temperature is not reached using the IPCC value for 200 years. Using Lewis and Curry's TCR estimate, it isn't reached at all.

The value of CS, inclusive of all feedbacks, is probably the most important unknown in the whole climate debate, yet we know little more about it today than we did when the Charney Report[155] was published in 1979. Happer points out that climate researchers have proposed more than 50 mechanisms for the poor model performance detailed in Chapter 5 (Happer, 2021b). He closes his discussion of CO_2 sensitivity with this:

> "The simplest interpretation of the discrepancy [between the models and nature] is that the net feedback is small and possibly even negative. Recent work by Harde indicates a doubling sensitivity of $S = 0.6$ K." (Harde 2014), (Happer, 2021b, p. 32)

The simplest reason for the discrepancy is that the doubling sensitivity (CS) computed by the model average ($\sim 3.2°C/2xCO_2$) is too high. So, Happer's contention is that the "doom and gloom" predictions of the IPCC community are the result of overestimating the sensitivity to CO_2 concentration. If we used a more reasonable estimate of climate sensitivity, one that matches what we see in nature, we may find that there is really no need to combat CO_2-caused global warming. Next, we examine the intersection of climate and politics.

---

[155] (Charney, et al., 1979)

Chapter 14: The Balance between Science and Politics

TheBestSchools

"There is also the question that is sometimes raised regarding the propriety of scientists' involving themselves to such an extent in political matters in the first place. Many feel that the danger here is twofold:

"Politicians may become lazy and abdicate their responsibility to educate themselves so they can understand the issues themselves.

"Scientists may compromise the integrity of scientific research itself, to the detriment of its proper functioning when controversial matters arise again in the future.

"On the other hand, we understand that scientists are citizens and as such have the right—perhaps the duty—to bring their special expertise to bear on matters of great public importance. These competing requirements—as scientists, to remain aloof from politics; as citizens, to become politically engaged—must create a difficult balancing act for conscientious scientists to negotiate, or so it seems to us.

"How have you attempted to walk this particular tightrope in your own work?" (Karoly, 2021a, p. 39)

Karoly answered the question as follows:

"I think that it is critically important for scientists to communicate the best available scientific evidence and interpretation of that evidence in matters of public interest, not only to scientists but also to the general public, to business and industry, and to government and politicians. Decision-making in all areas should be evidence-based, not based on opinions." (Karoly, 2021a, p. 39)

What should scientists do and not do?

Karoly defines "engagement in politics" as standing for election or actively campaigning for one political party. He does not believe that providing scientific advice and information to politicians makes science political. He adds that publicly pointing out the scientific limitations or advantages of policy proposals is not being political, as long as it is applied to all political parties. This seems fair, a scientist that only criticizes the proposals of one party, is being political by this standard.

185

Karoly has actively advised politicians since 2001 through the IPCC and as a member of the Australian Climate Change Authority. He has also presented evidence to several Australian Senate Committee hearings.

TheBestSchools also asked Happer a similar question in his interview. His answer:

> "I think that great damage has been done to the reputation of all of science by the global-warming frenzy of the past few decades. Twenty years ago, supposedly expert scientists solemnly declared that our children and grandchildren would not know what snow is [(Onians, 2000)]. A few weeks ago, Washington, DC, struggled to dig out of three feet of snow, a record in many locales [January 22&23, 2016].
>
> "In accepting his 2007 Nobel Peace Prize (also won by Yassir Arafat), Al Gore said the summer Arctic could be ice-free by 2013 due to CO_2 emissions. I invite readers to have a look at [Climate4you.com]. A few minutes of inspection of the 'sea ice' link will show that there has been no significant change in sea ice since 2007. With all due respect to Nobel Laureate Gore, there was plenty of summer ice in 2013." (Happer, 2021a, p. 54)

There is still plenty of ice in 2021. The Arctic ice is below normal, but within two standard deviations of the mean. The Antarctic ice is about average as I write this in May 2021, the middle of the Antarctic fall. According to the Rutgers University Global Snow Lab, autumn and winter 2020 snow cover in the Northern Hemisphere was a little below the 1972 to 2021 normal, but there is still plenty of snow.

Happer is convinced that politics, government bureaucrats, the news media and non-governmental environmental organizations have corrupted climate science. He believes the root of the problem is money. These groups "make a good living from the immense sums being squandered to address the non-existent threat of human-caused climate change." (Happer, 2021c, p. 2). According to Happer, there is no evidence that global GHG emissions will have a harmful effect on climate.

Karoly is concerned about the spread of misinformation. He thinks that political or ideological actors are spreading scientific misinformation to cast doubt about the dangers of climate change (Karoly, 2021a, p. 38). He cites the book *Merchants of Doubt*, by Naomi Oreskes and Erik Conway (Oreskes & Conway, 2010) as support for his view.

The *Merchants of Doubt* and ExxonKnew

On page 39 of his interview, Karoly comes very close to suggesting that his opinion regarding man-made climate change is fact and not a matter of opinion—a bit scary considering the lack of observations supporting his view.

In a difference of opinion or interpretation of data, where is the misinformation to which he refers?

Background: Naomi Oreskes

Naomi Oreskes is a professor of history at Harvard and an environmental activist. She was part of the "ExxonKnew" campaign organized by tobacco lawyers and the Union of Concerned Scientists. The campaign intended to use the legal strategy used in the big tobacco lawsuits to attack large fossil fuel companies. They needed to accomplish two things. First, they had to convince the public that the dangers of climate change were real and connected to fossil fuels. Second, they had to convince the public the fossil fuel companies knew about the dangers and withheld the information for profit. These were the key elements of the successful campaign against the tobacco companies. Oreskes and the others met to discuss this strategy in La Jolla, California in 2012 (May, 2020c, pp. 131-147).

The tobacco lawyers explained to the group that they won their case because it was already well established that tobacco causes cancer and cancer had a measurable cost to both governments and the public. Thus, there was a basis for computing a settlement. Then, during discovery they found evidence that the tobacco companies had knowingly lied about the dangers of smoking.

This caused a serious problem for Oreskes and the others. Seth Shulman reports that Claudia Tebaldi, a climate scientist at Climate Central, told the group:

> "If you want to have statistically significant results about what has already happened [on the health impacts of climate change]," she said, "we are far from being able to say anything definitive because the signal is so often overwhelmed by noise." Claudia Tebaldi (Shulman, 2012, p. 15)

Since damages due to fossil-fuel-caused climate change could not be established, they had a serious problem showing harm caused by Exxon and other large fossil-fuel companies. Their next goal was to show that "Exxon Knew" climate change was dangerous but hid it from the public and lied by omission. However, Exxon had published all their research on human-caused climate change and had serious internal disagreements over whether there was any danger. A thorough review of the published and unpublished Exxon research papers on the subject showed that they could speculate humans might be causing climate change but were unable to prove it because natural variability was as large as observed changes (May, 2016p).

Oreskes and colleagues published papers on climate change to lay a peer-reviewed foundation for the lawsuit that could be cited in court briefs. Then, finally in 2018, they had their day in court. Oreskes and Geoffrey Supran had published a "content analysis" paper that purported to show Exxon misled the public on their climate change findings. They claimed there was, "a discrepancy between what ExxonMobil's scientists and executives discussed about climate change privately and in academic circles and what was presented to the general public."[156]

Oreskes' analysis was destroyed in court by Kimberly Neuendorf, the leading expert on content analysis. From her court testimony:

> "I [Neuendorf] have concluded that S&O's [Supran and Oreskes] content analysis does not support the study's conclusions because of a variety of fundamental errors in their analysis. S&O's content analysis lacks reliability, validity, objectivity, generalizability, and replicability. These basic standards of scientific inquiry are vital for a proper content analysis, but they are not satisfied by the S&O study." (ExxonMobil, 2018a, pp. Attachment A, page 2)

Most of the errors identified by Neuendorf spring from poor sampling of ExxonMobil content. Supran and Oreskes improperly grouped together communications that vary across time and by author and audience. They also grouped together statements by Exxon and Mobil made prior to their merger as if they were one entity. Further, S&O coded the communications themselves rather than using objective and uninvolved coders, rendering their work non-replicable and unscientific. So, Oreskes, the Union of Concerned Scientists, and the tobacco lawyers could not prove any harm had been done by fossil fuels, nor could they show Exxon, or any other company, had lied. Their case fell apart and so did later cases filed by state attorneys-general (May, 2020c, pp. 141-147).

Most outlandish claims made by climate alarmists come from politicians, the news media, and other non-scientists. Examples include Oreskes, Al Gore, and Senator Sheldon Whitehouse. Because there is no observational basis for dangerous human-caused climate change, their claims harm their cause. Some scientists are reluctant to criticize these fervent advocates, which makes matters worse. Politics and science do not mix well, government corrupts science and vice-versa. It is a destructive combination with no benefits. Science only works when it is seeking for the truth.

We mentioned above that Karoly seems to support restricting the spread of what he calls misinformation. Karoly's misinformation, is Happer's truth. They have the same data but interpret it differently. Who judges what is misinformation and what is information? If there were a judge of the truth, do we lose freedom? We have free speech and a free press in the U.S. and Australia so that all views can be aired, and the public can sort it out. If some

---

[156] (ExxonMobil, 2018a, pp. Attachment A, page 2)

truth authority or government decides what is true and what is not, serious problems can arise. We talk about those problems next.

Chapter 15: The Bullying of Skeptics and Free Speech

TheBestSchools posed a question to Karoly regarding free speech:

> "[TheBestSchools has] a position on the subject of freedom of speech. The aspect of the whole [climate change] debate that troubles us the most is the very widespread notion that no one who questions the consensus can be doing so in good faith—that all critics of the consensus must be dancing to the tune of their paymasters in industry. It has gotten so bad that U.S. Senator Sheldon Whitehouse (D-RI) has recently called for the RICO Act against racketeering to be used to prosecute critics of the consensus! [see (May, 2020c, pp. 42-43 & 128-131) for more details]
>
> "But, after all, supporters of the consensus have a strong economic interest, as well—most of them are in the pay of some government bureaucracy! So, given this parity of 'interests,' it seems to us that both sides ought to put an end to the ad hominem attacks and focus instead on the actual evidence and arguments.
>
> "How do you feel about preserving scientists' and others' right to disagree on the issue of global warming?" (Karoly, 2021a, p. 37)

Karoly's answer was:

> "I strongly support the freedom of speech of scientists and of all people. Freedom of speech should not be used as an argument to support the dissemination of misinformation." (Karoly, 2021a, p. 37)

Opinions are opinions and not facts

Karoly then says that science is a process of repeatedly assessing evidence by subjecting it to the scrutiny of experts and that opinions are opinions, not facts. Further on, he says that the vast body of evidence, from many thousands of peer-reviewed studies, supports the consensus conclusions on global warming. He continues, saying:

> "In many jurisdictions, the deliberate spreading of misinformation for financial gain is illegal. Whether the deliberate spreading of misinformation about climate science to support the continued sale of fossil fuels for financial gain is illegal is a question that will need to be determined in the courts. It is fine for scientists and others to disagree on matters of opinion, but not on matters of the major scientific conclusions in climate science." (Karoly, 2021a, pp. 38-39)

Senators try to legislate a scientific outcome

Karoly seems to think that the "major scientific conclusions in climate science," which we read as the conclusions of the consensus, are facts and that disagreeing with them is the deliberate spreading of misinformation. Thus, he appears to support Senator Sheldon Whitehouse using the RICO statutes to arrest and punish climate skeptics because they disagree with the consensus. Steven Koonin reports in his book *Unsettled*, that "Senator [Charles] Schumer (together with Senators Carper, Reed, Van Hollen, Whitehouse, Markey, Schatz, Smith, Blumenthal, Shaheen, Booker, Stabenow, Klobuchar, Hassan, Merkley, and Feinstein) introduced Senate bill S.729." It says, in part:

> ". . . to prohibit the use of funds to Federal agencies to establish a panel, task force, advisory committee, or other effort to challenge the scientific consensus on climate change..."

Background: Legislating scientific truth

Yes, these Senators were trying to legislate against challenging an unproven scientific hypothesis, basically a scientific opinion, using government research money. It doesn't get much worse than that. Science is debate, if all sides are not examined and argued, it isn't science. Thankfully, their effort failed, the last thing we need is for scientific truth to be determined by the U.S. Senate!

When the climate data and model projections are as uncertain as they are today, and billions of dollars are spent on climate research every year, it would seem the consensus could be accused of spreading misinformation for financial gain. We agree that opinions are opinions and facts are facts, but Karoly and the Senators need to learn how to tell the difference.

TheBestSchools ask Happer a similar question:

> "[On] freedom of speech. ... the public conversation on global warming has taken on a decidedly nasty tone. Overt bullying and intimidation are now the order of the day. ...
>
> "One of the frequent charges (one you are not unfamiliar with personally) is that consensus critics are in the pay of the oil and gas industry. Such critics.... nearly all of whom work for government in one form or another.... do not often stop to consider that they, too, serve someone's economic interests ...
>
> "Given this 'parity' of interests between consensus critics and supporters, what do you make of the situation? Isn't the freedom to think what we like and say what we think at the very heart of the scientific endeavor?" (Happer, 2021a, p. 53)

Given that observations do not support the idea that humans control the climate with their fossil fuel emissions—or that climate change is dangerous—there is room for debate on the issue. And science is all about debate. In the western world, simply declaring one opinion is correct and suppressing opposing opinions is unacceptable.

Happer's response:

> "It is not possible to make progress in science without controversy. For example, there was heated scientific debate over Alfred Wegener's theory of continental drift for decades. ...
>
> "What is different about the global warming controversy is the direct involvement of governments on one side. As you mention, congressional demands that racketeering charges be brought against climate skeptics are unprecedented in the USA, ...
>
> "Bernie Sanders says he will 'bring climate deniers to justice' when he becomes President of the USA. What should people like me expect from President Sanders.... a concentration camp? The firing squad?
>
> "And what ever happened to the First Amendment to the US Constitution? ...
>
> "I would be surprised if the net total funding of climate skeptics exceeded $2 or $3 million dollars a year, and even that may be high. In the last few years, US government spending for climate research has been about $20 billion dollars a year.... more than a thousand times greater than skeptic funding. But even this huge financial advantage is not sufficient to support the pathetically weak scientific case that the world is in danger from more CO_2." (Happer, 2021a, pp. 53-54)

For those interested in the government climate change research spending details to which Happer refers, see *The Federal Climate Change Expenditures Report to Congress* (Office of the President, 2013). Many leaders of the Democratic Party call for the prosecution of people who disagree with the so-called consensus view. If they can find evidence that climate change—man-made or natural—is dangerous, they certainly have an audience and a platform. But, without evidence, the potential danger is just an opinion.

Debates like the one we describe in this book are the way to make a scientific case, if you have one, when the data are ambiguous. If you cannot make your case in a debate, imprisoning your opponents is not an acceptable alternative.

Happer attacked by Greenpeace

Greenpeace, and affiliated organizations, like the Climate Investigations Center and Our Next Economy, have attacked many climate skeptics (including Happer) throughout the years. Many of their attacks were detailed in our previous book (May, 2020c, pp. 49-86). The attacks follow a pattern. First, they identify their victim, and then link the victim to a fossil-fuel or energy company. The connection may be perfectly OK, but they spin it as nefarious. Then they feed the story, with as much negative spin as possible, to their numerous friends in the news media to build up the story.

Their news-media friends typically work for the *New York Times*, *Washington Post*, *PBS*, or *NPR*. These news sources have a very left-wing bias, and often report the "story" exactly as presented by the accusing organization, without looking into it deeply.

In Happer's case, a Greenpeace operative, Jonathon Ellis, pretending to be an agent for a Middle Eastern exploration and production "client," sent an email to Happer asking him to write about the benefits of additional CO_2 in the atmosphere. Carbon dioxide's benefits is one of Happer's favorite research topics and extremely important to him. Positive research on the subject—such as Idso's calculation of a three-trillion-dollar benefit to the world since 1961 from increasing atmospheric CO_2 and its effect of increasing the efficiency of photosynthesis in plants—are greatly under-reported. He uses every avenue possible to disseminate the good news. He even helped establish the CO_2 Coalition, a non-profit educational organization to inform policy makers and the public.

As a result, Happer was delighted to accept the offer. In response, he sent Ellis a paper he had co-written on the benefits of CO_2, mentioned his new CO_2 Coalition organization and said: "We would be glad to try to help if my views, outlined in the attachments, are in line with those of your client." (Happer, 2015). Find the complete email exchange between Happer and the Greenpeace mole online at (Happer, 2015).

Later in the email chain, Happer is careful to make sure Ellis and his client are clear about his views and his belief that there are real pollutants generated when fossil fuels are burned. These include oxides of sulfur and nitrogen, particulate matter, heavy metals, etc., but he does not believe CO_2 is one of them. He is clear that his work to push back on climate extremism is a labor of love, he wants no money for himself, but if a fee for his work is paid, it should go to the CO_2 Coalition. He pointed out to Ellis the arrangement was the same as he made with the Peabody Coal Company, now called Peabody Energy (NYSE: BTU).

Ellis, the Greenpeace operative, continued his deception until Happer testified before the U.S. Senate Subcommittee on Space, Science and Competitiveness on December 8, 2015. Then they fed the story of their subterfuge to the news media. Most in the news media ignored the lies told by the Greenpeace moles, did not check the facts, and told the story the way Greenpeace wanted it told.

The Guardian termed the operation a "sting" and said Greenpeace posed as fossil-fuel consultants and approached Happer asking him to write a report "touting" the benefits of rising CO_2 levels. Then added—without evidence—

that his views "are well outside mainstream climate science." The *New York Times* used the same terminology, also calling the Greenpeace deception a sting. The *Times* also reported "Greenpeace posed" as representatives of energy companies. The articles appeared on December 8 and 9 that year.[157]

Besides the obvious bias in both articles, they are also problematic for other reasons. Greenpeace and both newspapers imply that there is some sort of problem with fossil-fuel companies hiring researchers to write articles on their behalf. In the famous 2010 United States Supreme Court decision in Citizen's United (Hudson, 2017), the court clearly acknowledged that freedom of speech applies to corporations as well as individuals and that the government cannot restrict them from spending money to promote their ideas. Another law, 18 U.S.C. § 241, makes it a felony "for two or more persons to agree together to injure, threaten, or intimidate a person in any state, territory or district in the free exercise or enjoyment of any right or privilege secured to him/her by the Constitution or the laws of the United States." It seems that is exactly what Greenpeace was trying to do to William Happer, Peabody Energy, and any other fossil-fuel company choosing to hire Happer to write about the benefits of more CO_2.

Happer was willing to write the paper without any compensation, to get his message on the benefits of CO_2 out to the public, and to encourage a donation to his favorite organization, the CO_2 Coalition. This is his first amendment right. Greenpeace, with the enthusiastic cooperation of the *New York Times*, *The Guardian*, and other biased sources, attempted to spin his efforts as a negative. In so doing, they came close to breaking the law themselves.

The Guardian has no definition of "mainstream climate science" and offers no evidence Happer is "well outside" of it—whatever "it" is. Many well-known and accomplished scientists believe more atmospheric CO_2 is beneficial for life on Earth and have gathered a considerable amount of data to support the idea. Happer cites many peer-reviewed papers supporting his view, and we have added to his list in this book. The IPCC's AR5 downplays this concept, because they have an agenda, but it does acknowledge that CO_2 fertilization is real (IPCC, 2013, p. 484).

CO_2 clearly benefits life on Earth, we can measure the effect today.[158] The possible damages to Earth's ecosystem due to global warming and additional CO_2 cited by the IPCC, Karoly, and Tamblyn are all uncertain projections. The possible future damage they cite cannot be measured today. Models and projections are not observations whether they are mainstream science or not.

While additional CO_2 may be beneficial, it is a greenhouse gas and should cause some surface warming. We cannot measure the total effect of CO_2 on

---

[157] (Schwartz, 2015) (Goldenberg, 2015)
[158] (Zhu, Piao, & Myneni, 2016)

climate in nature because nature is too complex. But it is possible to compute the direct effect of additional CO_2 on surface temperature with a model. This assumes nothing else changes and there are no clouds, so it is not very realistic, but it can be done. Happer explains how in his Major Statement, and we try and explain it, with fewer equations, in the next chapter.

Chapter 16: Atmospheric Physics

All the debate participants agree that CO_2 is increasing in the atmosphere due to humans burning fossil fuels. Where they differ is on the impact. Happer believes the impact on our climate from the increasing CO_2 will be minimal and generally beneficial. Karoly, Tamblyn, and the IPCC believe that the additional CO_2 will lead to dangerous changes to the climate.

Unfortunately, we have no observations to guide us. If rising atmospheric CO_2 is warming the planet, it has stayed below natural variability, and has yet to be observed. Kevin Trenberth is among those that have written about this problem (Trenberth, 1991).

> "Natural variations, such as those associated with El Niños, make it difficult to detect anticipated [human-caused] climate changes. In addition, detection in the observational record of the expected global warming is confounded by flawed and patchy observations and because observed climate change is not geographically uniform. ... The observed patterns of temperature change over the globe are not yet well accounted for by climate models." (Trenberth, 1991)

How large is natural variability?

Recent work by Tongwen Wu and colleagues has attempted to untangle the multi-decadal Atlantic and Pacific climate cycles from the rest of the warming signal and then attribute the residual to CO_2.[159] They found that the AMO and PDO accounted for 30% of the warming over the past 140 years. However, they did not look at any other possible sources of natural variability, such as the solar wind, the solar magnetic field, or solar output. They also ignored longer term climate oscillations, such as the ~1,000 year Eddy Cycle (Eddy, 1976).

Ronan Connolly and 22 co-authors took a comprehensive look at the literature on solar variations as a cause of warming. They concluded that much of the warming in the Northern Hemisphere, and possibly all of it, may have been due to solar variability, not CO_2 (Connolly et al., 2021). Using plausible assumptions and estimates of the solar impact on climate Ronan Connolly and his colleagues found that the Sun could have provided anywhere from zero to 100% of the forcing necessary for recent warming. In short, the magnitude of natural variability is simply not known with sufficient precision.

Karoly recognizes that internal and solar natural variability must be understood to compute human influence on climate. He writes:

---

[159] (Wu, Hu, Gao, Zhang, & Meehl, 2019)

"To determine the likely causes of the observed increase in global average temperatures over the last century or the last 50 years, we have to assess the relative magnitudes of the contributions from natural chaotic variability of the climate system, from increasing greenhouse gases, from changing sunlight from the sun, and other possible factors." (Karoly, 2021a, p. 22)

In order to show that CO_2 is the major factor in atmospheric warming, he refers to the AR5 modeling results. Specifically, he refers to the diagram of climate forcings shown in Figure 27.[160] This assessment of the impact of CO_2 versus natural forces and internal variability is based on unvalidated models. The models have predicted much more warming than we have observed, as described in Chapter 5 and in AR6.[161] So, the reference is not helpful.

He also discusses the effect of variations in Earth's orbit around the Sun, usually called the Milankovitch cycles (Buis, 2020). Karoly's logic, as explained in his interview, is that CO_2 is going up and temperatures are going up. He does not think that, over the past 50 years, there have been any significant changes in the Sun, its magnetic field, or the solar wind. He also believes, as does the IPCC, that solar activity has declined over the past 50 years. Finally, he assumes that internal variability over the period has had no effect, see Figure 27, in Chapter 5, for the details. Connolly and colleagues (2021) list the various solar models that could explain much of the warming over that period. There are many, and all are ignored by Karoly and the IPCC.

Tamblyn offers more than IPCC references. He believes Happer is underestimating water vapor feedback, and writes:

"The Professor's view that climate sensitivity (CS) is quite low seems to derive from not understanding the water vapor feedback. As I discussed in my Response, this is a routinely observed feedback, part of normal meteorology. And satellite measurements observe a rising trend, a feedback." (Tamblyn, 2021b, p. 10).

Tamblyn then shows an IPCC plot of rising total water vapor over the oceans from 1990 to 2012. Carl Mears and colleagues published a similar satellite analysis in 2018 (Mears, et al., 2018). However, Rasmus Benestad (Benestad, 2016) and Ferenc Miskolczi (Miskolczi, 2010) found from weather reanalysis and weather balloon data, that global total water vapor has declined, although their work has been challenged.

---

[160] (Figure 11 in his interview) (Karoly, 2021a, p. 23)
[161] (IPCC, 2021, pp. 3-23 to 3-24).

Background: Total Precipitable Water

The jury is still out on the amount and direction of water vapor concentration in the atmosphere. Further, the impact of any change in the climate due to variations in water vapor is unknown, the signal is swamped by the effect of clouds. The impact of clouds could be positive or negative relative to surface temperature, we just don't know at this time, and cloud cover probably changes with water-vapor concentration, but even how this works is unclear. Tamblyn only succeeds in highlighting how little we know about water-vapor feedback. The RSS, ocean-only TPW is compared to the HadCRUT4 land-and-sea temperature anomaly and the National Centers for Environmental Prediction (NCEP) version 1 and 2 land-and-ocean TPW records in Figure 75.

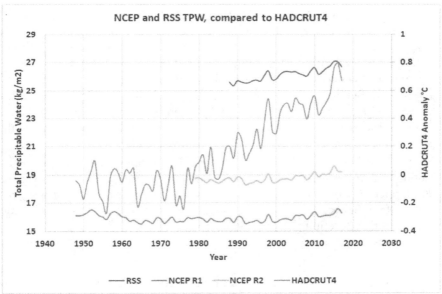

Figure 75. The RSS ocean-only TPW is compared to the HadCRUT4 land-and-ocean temperature anomalies and the NCEP version 1 & 2 weather reanalysis land-and-ocean TPW estimates. Source: (May, 2018d)

The NCEP weather reanalysis datasets, which use weather balloon, airplane, satellite, as well as land and ocean records show no definitive TPW trend until 2000, when they begin to trend upward. The HadCRUT global temperature rises quickly in comparison from 1975 to 2018. The RSS-satellite TPW dataset shows an upward trend from 1990 to 2018. Notice the NCEP version 1 and 2 estimates differ by nearly three kilograms, giving us an idea of the potential error. The RSS estimate is over the oceans only, which accounts for some of the difference between it and the other two. The NCEP R1 TPW estimates trend down to 1965, then remain flat to 2000, and trend up afterward. This is odd since the Pause in warming started around 2000.

Radiative Cooling of the Earth

While Karoly simply refers us to the IPCC and asserts that human greenhouse-gas emissions (mainly CO_2) are responsible for recent warming, Happer is not afraid to go into the details. He offers us a clear, but highly mathematical, picture of how energy warms Earth's surface and how Earth cools itself. The math can be difficult to follow, so we try to explain it in simpler terms. The reader should be cautious, mathematical expressions cannot be translated into words exactly. The true model is in the math, and in Happer's Major Statement—this section is only an approximation.

Karoly backed out of the debate after seeing Happer's explanation and did not respond to TheBestSchool's efforts to contact him. Tamblyn, Karoly's replacement, tries to refute it, so his comments are included, and we try to reconcile the two. It is a shame Karoly dropped out of the debate. Although Tamblyn was an appropriate replacement, it would have been a better and more robust debate had Karoly stayed with it.

As you may recall from Chapter 1, Happer is an expert in the interaction between energy and matter, especially between the atmosphere and radiation. It was this expertise that led him to invent the sodium-guide-star concept that is so critical for missile defense and astronomy today.

In 2020 Happer and William Wijngaarden wrote a paper in which they defined the radiative, clear sky, greenhouse effect very precisely (Wijngaarden & Happer, 2020). Much of the paper is an expansion of what Happer wrote in his Major Statement, which is discussed below. The 2020 paper is discussed in the section "Defining the Greenhouse Effect."

Energy Balance

Conduction of heat from the interior of the Earth is about 0.08 W/m², which is only 0.02% of the mean thermal energy from the Sun. So, according to Happer, virtually all the energy warming the Earth's surface is from the Sun, and we will focus on this energy and the ways that Earth's atmosphere and oceans work to keep the surface cool.

The concept of energy balance is critical to understanding climate science. It basically says that for the climate to remain somewhat stable, the outgoing energy from Earth must equal the incoming energy. Figure 76 is a graph of incoming energy from the Sun and outgoing energy to space.

The Sun radiates energy to Earth at a temperature of 5,525K (5,252°C) as illustrated with the red, theoretical curve. The temperature determines the frequency distribution, and within that frequency range, the peak frequencies are visible light. When solar energy hits our atmosphere, higher frequencies (shorter wavelengths) are attenuated by nitrogen, as can be seen in the major component box labeled Rayleigh Scattering in Figure 76, and to lesser extent by oxygen and ozone. Much of the remaining solar energy is attenuated by water vapor in the short-wave IR band of frequencies, creating the divots in the red spectrum of the spectral-intensity chart. In the absence of clouds, 70 to 75% of the incoming sunlight makes it through the atmosphere and strikes Earth's surface, where it is absorbed.

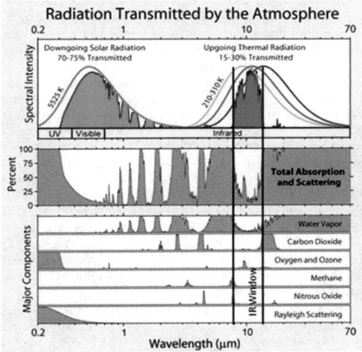

Figure 76. The energy arriving to Earth's surface from the Sun (left) ana the energy from Earth's surface transmitted to space (right) by frequency. From Happer's Major Statement, page 7, original author Robert Rohde, this figure is in the public domain.

Because most of Earth's surface is water, much of the solar energy striking it causes evaporation. When the Sun is directly overhead, on a clear day, in the ITCZ, the ocean surface balancing temperature is about 30-32°C, elsewhere it is less. This is a good thing, since the lower atmosphere is opaque to most OLR from the surface, so radiating away the absorbed solar energy is not an option, it won't get very far before it is absorbed by a GHG. The evaporation raises the humidity of the overlying air. Water vapor is less dense than dry air, so it rises, initiating convection which increases the rate of heat transfer.

As the humid air rises, it expands and cools until it reaches an altitude where it is cool enough for the vapor to condense into droplets and form clouds. This can happen quietly, or violently as a thunderstorm, but this cooling process occurs in every part of the Earth where it is warm enough and where the surface has sufficient water.

Clouds make a huge difference in climate and, unfortunately, the cloud forming process is so complex and happens at such a small scale it currently cannot be modeled. Clouds are generally white and reflect a great deal of sunlight. They are responsible for much of Earth's albedo (reflectivity). Clouds not only reflect solar short-wave radiation to space, they also emit heat from condensing water vapor and absorb and reemit thermal energy transmitted from the surface. So, they work to reduce the solar energy striking the surface, and simultaneously warm the surrounding air and surface—the latter mostly at night. Their net impact on surface temperatures changes with the time of day, the location, and the type and altitude of the cloud.

Radiation reaching the Earth from the Sun falls within the ultraviolet, visible-light, and SW bands, upward radiation emitted from Earth to space falls within the OLR bands (right side of Figure 76). While the Sun emits at 5,525K, the Earth emits its radiation at a much lower temperature—mostly between 210 and 310K (-63° to 37°C). Some Earth emitted surface radiation can go directly to space with the lowest temperatures emanating from Antarctica, and the highest from the Sahara and other deserts, accounting for the radiation as warm, or warmer than 37°C.

Except within the frequency range of the IR Window (or infrared window) noted in Figure 76, outgoing radiation to space is mostly blocked nearly everywhere, by water vapor. This means a relatively large portion of outgoing radiation, from the surface, comes from the drier poles and deserts where humidity is low. CO_2 also blocks a small, but significant, portion of the frequencies plotted in the graph.

In summary, about 10 to 30% of the energy emitted from Earth's surface makes it directly to space, mostly through the IR Window. Nearly all the rest is carried away from the surface through evaporation of water, ocean and wind currents, or sensible heat transfer. The heat, carried by water vapor and warm air, moves through the atmosphere via convection and sensible heat transfer until it reaches an altitude where it can be radiated to space. This altitude is largely determined by the local cloud-top height, since above the tops of the local clouds the humidity is much lower. Higher in the atmosphere, where humidity is essentially nonexistent, much of the transmission of thermal energy to space is done by CO_2.

Determining the temperature of the Earth's surface

Happer begins his discussion by telling us that if the Earth had no Sun, it would be very cold. The Earth's interior is warm, but only 0.08 Watts per square meter (W/m²) of thermal energy makes it to the surface. In climate studies, this is called "forcing," and the common units for forcing are W/m². Next Happer introduces the Stefan-Boltzmann equation shown in Figure 77.

In this equation, J is the forcing in Watts/m², ε is the emissivity, σ is the Stefan-Boltzmann (SB) constant (5.67 x 10⁻⁸) and T is temperature in Kelvin. Figure 78 is a plot of this equation from zero Kelvin (-273.15°C) to 330 Kelvin (57.15°C).

$$J = \varepsilon \sigma T^4$$

Figure 77. Stefan-Boltzmann equation.

The SB equation applies to a theoretical construct, called a "blackbody." A perfect blackbody is opaque and reflects no light or energy. It absorbs all radiation and then re-emits all of it in frequencies that depend upon its temperature, which is assumed to remain constant. A pure blackbody has an emissivity and absorptivity of one. This is generally visualized as a spherical black cavity that has a small opening to allow energy in and out. The cavity is held at a constant temperature, so regardless of the frequency of the incoming radiation, the radiation emitted by the interior is dependent only on the temperature of the cavity. No energy is reflected, and the body is in thermal equilibrium, so the emissivity equals absorptivity according to Kirchhoff's law.[162]

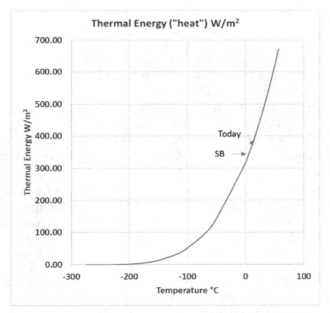

Figure 78. Plot of temperature vs. forcing. Identified are both today's global average temperature of 15°C and Happer's calculation of the Stefan-Boltzmann black-body temperature of 5°C.

---

[162] (ERK, 2010)

Obviously, the real Earth is not a perfect blackbody and reflects about 30% of the solar radiation that strikes it, as Happer writes in his Major Statement (Happer, 2021b, p. 10). Most of the radiation reflected is from clouds, ice, deserts, or water.[163] Water is not very reflective, as it has a low albedo, but it covers over 70% of Earth's surface. Emissivity and absorptivity are complex in real life and depend on many constantly changing factors. However, Happer is not talking about the real Earth, he is demonstrating an important physical concept.

Likewise, surface temperature is complex and depends upon altitude and latitude, among other factors. But Happer simplifies the calculation and assumes that emissivity is one. Using the equation in Figure 77, we find that the 0.08 W/m² from the Earth's interior results in a surface temperature of 34K or –239°C: very cold. We need the Sun!

Next Happer asks what is Earth's temperature, if it radiates the same energy as it receives from the Sun, like a blackbody? The Sun delivers 340 W/m² and the SB equation tells us this results in a temperature of five degrees Celsius (marked as SB in Figure 79). This is close to today's average surface temperature of about 15°C.

Tamblyn objects to this mathematical exercise and writes the following in his Detailed Response:

> "Professor Happer calculated the expected average temperature of the Earth using the Stefan-Boltzmann equation, arriving at a figure of 5°C. However, his calculation contains a *significant error*: He doesn't account for the 30% of sunlight that is reflected and is not absorbed! The Earth does not need to radiate as much as his calculation suggests and when we redo the calculation accounting for this, the average temperature the Earth needs to be at to balance incoming solar radiation is −18°C.
>
> "But the surface is at around 15°C! The surface of the Earth is over 30°C warmer than we might expect due to simple energy balance alone, not the 9°C Professor Happer calculated. The greenhouse effect explains this." (Tamblyn, 2021a, pp. 4-5)

Tamblyn did not read what Happer wrote—or didn't understand it. The normal theoretical calculation of Earth's blackbody temperature is –18°C to -19°C since it subtracts the radiation Earth reflects, as Tamblyn says. But this is an artificial calculation. The Earth is not a blackbody, and its emissivity and absorptivity are much less than one, critical points that flew right over Tamblyn's head. A blackbody is in thermal equilibrium by definition and has

---

[163] (Coakley, 2003)

a constant temperature. The Earth's atmosphere is never in thermal equilibrium, particularly at the surface where convection is constant.

The calculation that Tamblyn mentions is for a blackbody with an assumed emissivity of one, after reflected energy is subtracted. For this to be true the atmosphere, ice caps, and oceans would have to disappear, and the Earth would have to have a uniform coating of some perfect material that would absorb all solar radiation and quickly re-emit it. This is contradictory and does not account for Earth's constantly changing albedo or reflectivity, the reflectivity subtracted is an average. Happer made no error, he stated exactly where his calculation came from and the assumptions he made. Happer's response, from his Final Reply, is appropriate:

> "I did not pretend to a precise calculation of the average temperature of the Earth with its atmosphere and clouds. My approach was the classic one that has served physics so well over the centuries: start with the simplest model and then improve it." (Happer, 2021d, p. 3)

Unfortunately, many of Happer's arguments were treated in a similar fashion by Tamblyn. This reduced the quality of the debate but is illustrative of what has been happening in the general news media. The media and the global warming consensus often simply assume that skeptics, like Professor Happer, don't know what they are talking about. They dismiss their arguments, or pick at some trivial inconsistency, without thinking or doing their homework. This attitude is more akin to politics than science.

Atmospheric transmission of radiation

The structure of our atmosphere is shown in Figure 79. The lower level of the atmosphere—the troposphere—is where weather and convection occur. The top of troposphere is around 11 km in the mid-latitudes, as low as six km at the poles and as high as 17 to 18 km at the equator. In the troposphere most heat transfer takes place via convection (wind), evaporation, and ocean currents.

The troposphere cools with altitude, the rate of cooling varies greatly with location and weather, but on average, the temperature drops 6.5°C per kilometer, as shown in Figure 79. Above the tropopause, in the stratosphere—where there is little water vapor, circulation, or convection—the trend reverses, and temperature rises with altitude. The stratosphere is warmed when solar radiation is absorbed by ozone (O_3). The stratosphere is mostly cooled via radiation to space, much of it emitted by CO_2. More CO_2 will cool the stratosphere, just as more solar radiation warms it, if the ozone concentration, solar output, and all other factors stay the same. Wijngaarden and Happer estimate doubling CO_2 to 800 ppm will cool the stratosphere by 10°C.[164]

No one knows how much warming of the troposphere will be caused by an increase in CO_2. Estimates are all over the place. But Happer believes that the lower troposphere will warm about one degree if CO_2 doubles, and the middle troposphere will warm by 1.2 degrees. He is convinced the higher sensitivities published by the IPCC are a result of overestimating the water vapor and cloud feedback to CO_2-caused warming (Happer, 2021a, p. 18).

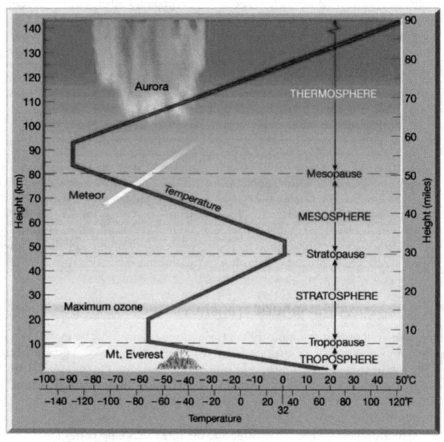

Figure 79. Atmospheric temperature profile. Source: (Faesi, 2013).

Figure 80 is the famous Blue Marble photograph of Earth, taken by Harrison Schmitt from Apollo 17. This photo was taken at mid-summer in the Southern Hemisphere and the ITCZ string of clouds is crossing southern Africa and running out through the Indian Ocean, north of Madagascar. As mentioned previously, it is here that ocean evaporation balances the thermal

---

[164] (Wijngaarden & Happer, 2020)

energy absorbed from the Sun on a clear day at about 30°C. There are not a lot of clear days in the ITCZ. The rising moist air in this zone pulls in humid air from all around forming heavy clouds that are present nearly all the time. So many, that the zone is easily identified from space. It is Earth's natural air conditioner at work. The ITCZ migrates north to south across the equator following the Sun's position throughout the year.

Clouds and ice provide a high proportion of Earth's albedo. In many ways clouds determine our climate, but their precise impact is unknown. And, once again, we don't even know if they work to increase or decrease surface warming.

Figure 81 shows the Western Hemisphere, but it looks at the IR window (Figure 76) frequencies. These frequencies can go all the way from Earth's surface to space, on a clear day or night, without being captured by water vapor or CO_2.

Figure 80. The Blue Marble Earth photo, taken by Harrison Schmitt from Apollo 17.

Now we see why clouds are so important. During the day, cloud tops reflect a great deal of sunlight, cooling the surface. But at night, clouds can block the transmission of infrared radiation from Earth's surface to space, warming the surface. This happens sometimes, even in the IR window.

The thickness of clouds, where they are located, and their altitude affect their impact. Thick clouds at the equator can reflect a great deal of solar radiation, thinner clouds allow a lot through. At higher latitudes, where the Sun is at an angle, even thin clouds can block a lot of incoming radiation.

Figure 81. Photo using emissions in the infrared window identified in Figure 76. Darker colors are heavier IR emissions and lighter colors are less.

The amount of radiation emitted is a function of temperature as can be seen in Figure 76 and the SB equation in Figure 77. The SB equation works both ways, higher energy causes higher temperatures, and higher temperatures cause the emission of more energy. When clouds are lower, they are warmer, and closer to the surface temperature. Thus, when lower clouds emit thermal radiation, they emit nearly the same amount as the surface. Since they reflect a lot of energy during the day and emit nearly the same amount of energy at night as the surface, they probably have a net cooling effect. However, the uncertainty in the effect of low-level clouds is very high and growing.

According to AR6, the overall low-level cloud impact uncertainty is higher in AR6 than it was in AR5.[165] We will see in the next section that Lindzen and colleagues have shown that higher-level cirrus clouds have a net warming effect in the tropics.

Lindzen's Iris Effect

As Happer explains in his major statement, Richard Lindzen and Yong-Sang Choi showed that changes in the extent of high-level cirrus clouds can be caused by heating or cooling of the surface in the tropics. High-level cirrus clouds consist almost entirely of ice crystals. The saturation vapor pressure of ice is very low, so cirrus clouds are persistent and cover large areas. Their daily variability alone, is measured in millions of square meters (Lindzen & Choi, 2021).

Background: the Iris Effect
Most of the water vapor in the atmosphere resides in the tropics. If we define the tropics as from 30°S to 30°N, then the tropics occupy 50% of Earth's surface, contain most of the world ocean, and—as shown in Figure 82—63% of the water vapor. This is also where temperatures are the warmest, evaporation is highest, and the troposphere is the thickest (up to 18 km).

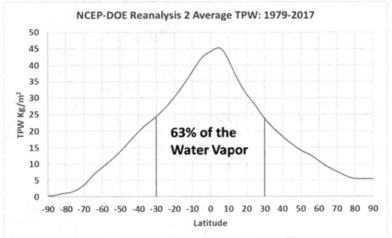

Figure 82. Total precipitable water in Kg/m², by latitude slices from the South Pole (–90) to the North Pole (90). 63% of the water vapor is in the tropics between 30° and -30°. Data from NCEP-DOE Reanalysis 2.

Thus, if water vapor is the feedback mechanism that increases CO_2-caused warming, this is the region where most of that feedback comes from—where the action is. As we discussed in Chapter 3, Christy and McKitrick chose the upper troposphere in the tropics for their statistical test of the climate models

---

[165] (IPCC, 2021, pp. 7-113 to 7-114)

and the models failed. CO_2 alone, without any positive feedback, is not threatening. The reason the models failed the test may very well be that they overestimated water-vapor feedback.

Lindzen shows that higher-level cirrus clouds are nearly opaque to outgoing IR radiation, and they are extremely high and very cold, so they do not emit much energy themselves. Satellite and surface observations suggest that incoming solar radiation is not significantly reflected by cirrus clouds, especially in the tropics, causing a net warming effect. From Lindzen's 2001 paper:

> "Essentially, the cloudy–moist region appears to act as an infrared adaptive iris that opens up and closes down the regions free of upper-level clouds, which more effectively permit infrared cooling, in such a manner as to resist changes in tropical surface temperature." (Lindzen, Chou, & Hou, 2001)

Lindzen and Choi found that these high-level clouds thin and separate, like the iris of an eye, as tropical sea-surface temperatures increase. So, they named the process the "iris effect."[166] In the Pacific tropics they found a strong correlation between cirrus cloud cover and SST. Cirrus-cloud cover decreases 22% for each degree Celsius increase in surface temperature.

They computed the climate sensitivity, including cloud feedback, to be between 0.5 and 1.3°C/2xCO$_2$. From the 2011 paper:

> "We again find that the outgoing radiation resulting from SST fluctuations exceeds the zero-feedback response thus implying negative feedback. In contrast to this, the calculated TOA outgoing radiation fluxes from 11 atmospheric models forced by the observed SST are less than the zero-feedback response, consistent with the positive feedbacks that characterize these models. The results imply that the models are exaggerating climate sensitivity." (Lindzen & Choi, 2011)

Lindzen writes the following for the CO_2 Coalition:

> "The situation with respect to climate sensitivity is that we basically see no reason to expect high [climate] sensitivity [to CO_2]. The original basis for considering that high sensitivity is possible (namely, the hypothetical water vapor feedback of Manabe and Wetherald, 1975) is clearly contradicted by the measurements of TOA radiative fluxes which show that the total long-wave feedback, including cirrus cloud variations, may even be negative. Analysis of the temperature data [shows] climate sensitivity in excess 1.5°C is precluded." (Lindzen, 2019, p. 21).

---

166 (Lindzen, Chou, & Hou, 2001) (Lindzen & Choi, 2011)

Lindzen and Choi initially focused on the tropics, but when they expanded their study to the entire globe, they produced similar results. However, the noise in the satellite signals they were using increased. They are of the opinion that the water-vapor feedback is primarily in the tropics, where most of the water vapor is.

Lindzen and colleagues hypothesized that the reduction in area of high cirrus clouds with SST was due to an increase in cloud water with higher SSTs. As cloud water increases, the formation of rainfall in convective clouds becomes more efficient. The effect is less water-vapor condensate is available for detrainment to form thin cirrus clouds in the upper troposphere (Lindzen, Chou, & Hou, 2001).

In other words, the iris effect constrains surface temperatures in the tropics, and this constrains climate sensitivity. Roberto Rondanelli wrote a paper[167] with Lindzen where they showed that convective precipitation increases with SST at a rate of 6–12%/°C. More recently, Lindzen and Choi wrote a review paper on the iris effect (Lindzen & Choi, 2021). In this paper, they point out that a strong reduction of high-level cirrus clouds appears in both model and satellite studies as a response to surface warming. Thus, the iris effect is well established. Figure 83 shows the three distinct tropical regions involved in the iris effect.

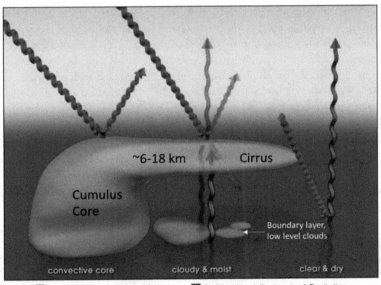

Figure 83. Diagram of a cumulus core and spreading cirrus. Diagram after NASA Earth Observatory (Lin, 2002).

---

[167] (Rondanelli & Lindzen, 2008)

First, there is a cumulus tower that takes water vapor extremely high in the troposphere. This region is relatively small and often called the "cumulus core." Next, spilling (or detraining) out of the tower, above 6 km, are cirrus clouds of ice crystals, this is the "cloudy-moist" region. Finally, we have the dry and clear area, where downward moving dry air keeps the humidity low, it lies beyond the cirrus clouds.

The relative areal sizes of each of these tropical environments are important in determining the magnitude and sign of water-vapor feedback. Normally climate models assume that relative humidity will remain constant, but studies of satellite data by Lindzen,[168] Spencer,[169] and many others show that this oversimplifies the picture. The water vapor response to warming is still unknown.

However, a warm surface does stimulate the production of the cumulus (or cumulonimbus type) cloud cores. Detrainment from such core clouds is the origin of about half of all cirrus clouds, the other half are formed from larger-scale dynamics lifting air to high altitudes (Lindzen & Choi, 2021).

The cirrus cloud level is well above the boundary layer, anywhere from six to 18 km. The boundary layer is often marked by the base of low-level clouds. The normal decrease of temperature with height reverses at the base of the local clouds, since it is where vapor condenses and begins to form water droplets, releasing latent heat and warming the surrounding air.

As we can see in Figure 83, cumulus clouds are thick and effective at reflecting incoming solar radiation. However, cirrus reflect some incoming radiation, but also let some through. Likewise, they reflect or absorb some upwelling radiation from the surface and low-level clouds. Low-level clouds are major emitters of IR radiation, cirrus clouds are much colder and emit much less. With no clouds and low humidity, thermal radiation makes its way to space quickly from clear and dry areas.

Everyone who has studied the issues agrees that the iris effect exists, it occurs in models and can be observed by satellite. Where opinions differ is on how it affects temperature. Lindzen and colleagues think it is a net-negative feedback, decreasing the CO_2-warming effect. Others think the iris effect enhances CO_2 warming, or that the warming and cooling cancel each other out (Lindzen & Choi, 2021). What does the evidence say?

Besides the ERBE and CERES satellite data, the Tropical Rainfall Measuring Mission as well as ground-based radar confirm that rainfall efficiency is improved and that high-level cirrus detrainment from cumulus towers (anvil clouds) decreases at warmer surface temperatures. That part of the iris theory is solid, but the exact magnitude of the effect is unknown. Also, the reason why rainfall precipitation increases when the surface warms is still not understood.

---

[168] (Sun & Lindzen, 1993)
[169] (Spencer & Braswell, 1997)

Numerous papers have been written criticizing the theory behind the iris effect, but none have held up. Most studies find that cloud and water-vapor feedbacks are inextricably mixed. Lindzen and colleagues recommend that rather than studying these feedbacks separately, it is better to study infrared and shortwave feedbacks, because they are easier to measure. Local measurements of cloud cover and humidity bear little relationship to global averages, making them poor measures from a global climate perspective. On the other hand, incoming and outgoing radiation can be measured from satellites over nearly the entire planet.

Lindzen and colleagues were able to establish a solid negative-longwave-feedback. Their shortwave feedback results were more ambiguous, but all their total feedback results were much smaller than those predicted by the IPCC climate models.

Thorsten Mauritsen and Bjorn Stevens acknowledge that convection efficiency is only very crudely represented in current climate models. They programmed a very simple representation of the observed precipitation efficiency/iris effect into their climate model and found this caused the computed ECS to fall to the low end of the IPCC range, near a value of two. They think the models might be missing these important hydrological feedbacks and agree with Lindzen that dry and clear areas in the tropics (see Figure 83) expand with surface warming (Mauritsen & Stevens, 2015). Mauritsen and Stevens showed that including the iris effect in their model moved all model results closer to observations.

It is unlikely that the iris-effect feedback is positive as proposed by Bing Lin (Lin, 2002) or neutral as proposed by Mark Zelinka (Zelinka & Hartmann, 2011). As Lindzen and colleagues point out, changes in radiative flux at the top of the atmosphere have a nearly simultaneous effect on SST, there is little to no lag. Many studies suggesting a positive or neutral feedback use gridded data that is lagged, which weakens the correlation.

Even when using non-lagged data, the signal is noisy, as Lindzen readily admits. The longwave-negative-feedback signal seems reliable, but the shortwave feedback may be positive, in terms of surface temperature. The net feedback is still unclear, but very likely negative or cooling.

Dennis Hartmann and Marc Michelsen were among the first critics of the iris effect (Hartmann & Michelsen, 2002). Many of their criticisms were either irrelevant or quantitatively trivial, and some were simply wrong (Lindzen, Chou, & Hou, 2002). The editors of *BAMS*,[170] included this statement above their article: "Careful analysis of data reveals *no* shinkage of tropical cloud anvil area with increasing SST." But, it turns out their criticisms of the iris effect and Lindzen and Choi's work had no merit, a major *BAMS* failure. Their main argument was that cloud-cover changes were due to clouds moving into the study area that were unrelated to cumulus convective cores. However, the data

---

[170] *Bulletin of the American Meteorological Society*

did not support the idea. Hartmann and Michelson emphasized that cirrus clouds did not need to come from cumulus cores, they could form in other ways, and ignored the procedure used by Lindzen and colleagues to make sure they only worked with cirrus clouds coming from cumulus cores. All the critiques of the iris-effect theory have been rebutted, and, to date, the rebuttals have withstood the test of time.

Neither Tamblyn, nor Karoly address Lindzen's excellent work in their documents, despite Happer's mention of it. Their opinion of the iris effect is unknown to us, but the effect has been well established in the literature for twenty years, is unrefuted, and fits observations. The exact value of the net feedback is unknown, but likely negative. AR6 estimates that the net feedback from the iris effect is about -0.13 W/m²/°C, but the precise value is highly uncertain (IPCC, 2021, p. 7-66). If nothing else, it does suggest a better test of current climate models. They should be tested on how accurately they predict incoming solar and outgoing infrared radiation, things that can be measured with reasonable accuracy from satellites. Predictions of SST and percent cloud cover are too ambiguous. In many ways, the IPCC and the consensus are looking into the wrong end of the telescope.

Planck's Function

Happer's description of his model of radiation from Earth's surface begins with the Planck function, shown in Figure 84. This function relates temperature and radiation frequency to power. Power can also be called brightness or radiance. Figure 81 shows where surface IR makes it to space and where it doesn't. The darker shading shows more radiation is exiting through the IR window, a large B. The lighter areas are where clouds are blocking (either reflecting or absorbing) some of the radiation emitted from the surface, a lower B.

The smooth dashed lines In Figure 85 are the theoretical blackbody-brightness functions derived from Max Planck's function:

$$B = \frac{h_p c^2 v^3}{e^x - 1}$$

Figure 84. Planck's function. This is plottea with the smooth lines in Figure 85.

The function relates spatial frequency v to brightness B or power per unit area, h_p is Planck's constant (6.63×10^{-34} Js), c is the speed of light, and e is the exponential constant. $x = h_p c v / (k_B T)$, which is the ratio of the energy, $h_p c v$, of a photon of spatial frequency v to the characteristic thermal energy $k_B T$; where

k_B = Boltzmann's constant = 1.38×10^{-23} J/K. T is temperature in Kelvin. The units of B are W/(m²·sr·cm⁻¹). Here, sr = steradian, the unit of solid angle.

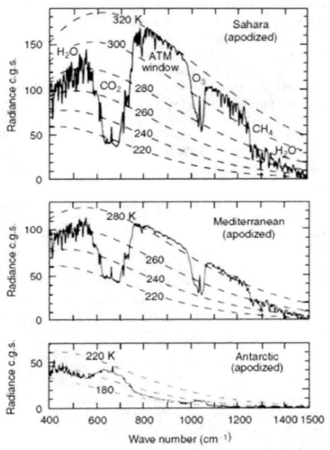

Figure 85. Outgoing IR spectra for the Sahara, Mediterranean, and Antarctic. The horizontal scale is the "wave number", the inverse of wavelength.

Kelvin	Celsius	Fahrenheit
320	46.85	116.33
300	26.85	80.33
280	6.85	44.33
260	-13.15	8.33
240	-33.15	-27.67
220	-53.15	-63.67
200	-73.15	-99.67
180	-93.15	-135.67
160	-113.15	-171.67

Table 1. Temperature conversions for Figure 85.

214

There are 4π steradians for all solid angles emanating from a point in 3-dimensional space. Thus, B is the power per square meter, corrected for the 3D emission angle it takes to space.

The graphs in Figure 85, show the frequency distribution of outgoing IR radiation for the Sahara Desert, the Mediterranean, and Antarctica. The vertical scale is radiance or brightness, the higher the value the brighter for a given radiation frequency. The pattern shown for each area is as it would be seen by a satellite on a cloudless day.

In Figure 85, Happer's Figure 8, the units on the Y axis are c.g.s., which are $erg/(s\ cm^2\ sr\ cm^{-1})$ or $mW/(m^2\ sr\ cm^{-1})$. Basically, the vertical units are power per unit area, but they are for outgoing radiation, so the units are complicated. We use the units W/m^2 elsewhere in the book and c.g.s. are similar but describe radiation moving in the opposite direction (Happer, 2021b, p. 14).

The spectra for the three areas are overlain with the theoretical "Planck-brightness" blackbody power versus frequency curves for the temperatures given in the figure in Kelvin. These curves are computed with the equation in Figure 84. Table 1 translates Kelvin (K) to degrees Celsius and degrees Fahrenheit for convenience. The Stefan-Boltzmann energy, as computed from the equation in Figure 77, is the area under the Planck brightness curves.

In the Sahara Desert, the IR Window (labeled ATM and short for atmosphere) shows energy reaching the satellite with an apparent blackbody temperature of 320K or 47°C or 116°F as shown in Table 1. This temperature suggests a hot desert.

The Mediterranean temperature is about 285K, or a comfortable 12°C. The Antarctic window is about 210K (−63°C). The measured data in Figure 85 is "apodized" meaning filtered to make the curves clearer and easier to read. In areas outside the IR window, the temperature of emission is lowered because the radiation was emitted at a higher altitude, where the air is cooler.

The large divot taken out of the Sahara and Mediterranean curves and labeled CO_2 in Figure 85, are a set of frequencies that are entirely blocked by the troposphere. The temperature where that CO_2-frequency energy is emitted, 220K, occurs in the tropopause and higher in the atmosphere. Some of these frequencies are emitted from as high as 84 km. Almost no water vapor or weather exists in this altitude range. This zone is cooled via radiation to space, nearly all of which comes from CO_2.

Most of the other deviations from the blackbody-surface-temperature curve result from radiation emitted by water vapor at about 260 to 280K. These temperatures occur between five to eight km in altitude—dependent mostly upon latitude—and above most clouds. Cloud tops are where rain and snow form and they are significant emitters of IR in the water vapor frequencies. Much of the energy is released when the vapor condenses into rain or snow. Some of the energy warms the surrounding air and much of the remainder goes straight into outer space or back to the surface.

In Antarctica, the CO_2 emissions come from a higher temperature than the ground. These emissions cool, rather than warm the air. In Antarctica, the ground is so cold, the prevailing winds bring in warmer air from the Southern Ocean and this forms a persistent temperature inversion. As a result, Antarctica has a negative greenhouse effect.

The Schwarzschild Equation

Happer then discusses the Schwarzschild equation, which is not reproduced here, but is shown in Happer's Major statement on page 16. The equation describes how the intensity of upwelling radiation changes with altitude, whether the source is the surface or GHGs. The brightness described is the radiation as seen from a satellite in space.

The equation shows us that the radiation emitted from an object decreases as the object's temperature drops. In the troposphere, where water and water vapor emit most of the energy, temperature decreases with height (see Figure 79), so the higher you go, either in altitude or latitude, the less energy (or power) is emitted by each H_2O molecule.

Next, we see that the atmospheric molecular density decreases rapidly with altitude. CO_2 is well mixed in the atmosphere and exists everywhere with almost the same atmospheric fraction. But water vapor is not, as it condenses to water at low temperatures. Methane and ozone are strongly reactive and are also poorly mixed. Methane and ozone are greenhouse gases but have such a small potential impact on climate we omit discussion of them here. Ozone warms the stratosphere but does so at a high altitude (>20km) and has little effect on the surface.

Water vapor is the strongest greenhouse gas but is nearly absent from the atmosphere in the polar regions and above the tropopause. When water vapor is absent, CO_2 is, by far, the strongest greenhouse gas.

There are several things that the Schwarzschild equation balances. The first is that the atmosphere is trying to make the local brightness (energy emission) equal to the Planck brightness curves seen in Figure 85. Secondly, as the radiation moves upward through the troposphere toward the satellite, it loses brightness, and that energy is lost to air molecules, warming the air. This reverses in the stratosphere, where the air gets warmer with altitude. In the stratosphere, the molecular emission intensity increases and the growth in energy comes from cooling the surrounding air.

Finally, the atmosphere is never in equilibrium, and it is not a blackbody. But, because the molecules in the atmosphere collide with one another billions of times a second, the overall energy state, at any location, is well described by the local temperature. The equation describes the transmission of energy from the surface to space and shows how the GHGs operate to warm the troposphere and cool the stratosphere. This is the "greenhouse effect" in a nutshell.

Where water vapor exists, it dominates the greenhouse effect and CO_2 plays a minor role. Where it does not exist, or exists in very small quantities, such as in the stratosphere, Antarctica, or in deserts, CO_2 plays a dominant role.

Logarithmic forcing by CO_2

The atmospheric window, labeled "IR window" in Figure 86 are frequencies that can travel from Earth's surface to space without being captured by GHGs. These frequencies, when measured by a satellite on a clear day, show the surface temperature. Comparing the IR window to the CO_2 frequency divot to its left, we can see that CO_2 is blocking nearly all surface radiation in the divot frequencies. Thus the troposphere is nearly opaque to radiation in the CO_2-frequency range. We know this because the CO_2 frequency divot reflects an emission temperature of 220K (-53°C), a temperature in the lower stratosphere or higher (see Figure 80). This is the lower stratosphere temperature over much of the Earth, the lower stratosphere is colder over the tropics, as low as 190K or -83°C and much higher. You can see in Figure 86, that even in Antarctica, where the surface temperature is lower, the emission temperature rises to 200K.

Over most of the planet, though, CO_2 warms the surface. The warming can be countered by clouds, precipitation, and convection, but outside of the polar regions, we do expect some warming from additional CO_2. How much warming?

Figure 86 is a simplified theoretical portrayal of Figure 85 for CO_2 only. The equations needed to make this plot can be found in Happer's Major Statement from pages 18 to 22.

The units used on both axes are same as in Figure 85. However, the y-axis is now labeled intensity. Recall that surface-emitted IR radiation can only reach space (the satellite) in the IR frequency window between about 800 to 1000 cm^{-1}. In this window, the surface radiation intensity (red curve) matches the Planck-function blackbody temperature (320K in the Sahara, 290K in the Mediterranean, and 180K in the Antarctic). Since the troposphere is opaque to IR, the divot in the curve between 600 and 730 cm^{-1} reflects a temperature above the troposphere—about 200K to 220K—depending upon the location (blue curve). This temperature is warmer than the surface in Antarctica, but cooler throughout much of the world.

Because the troposphere is already opaque to CO_2 frequencies, adding more CO_2 will have little effect. The solid black curves represent the effect of 400 ppm CO_2, the concentration we have today. The dashed black line represents 800 ppm CO_2. Most places it overlays the 400-ppm line, but at the edges of the CO_2 divot, we see the width of the 800-ppm curve is just a little broader than the 400-ppm curve.

Happer shows us how to calculate the broadening of the divot. For each region, two black dots are present along the 400-ppm line at 582 cm^{-1} and 756 cm^{-1}. He calculates the difference for both 400 ppm and for 800 ppm along the horizontal axis, and then the difference in the transmission to space. He finds that broadening the divot decreases the radiation to space by about four W/m^2.

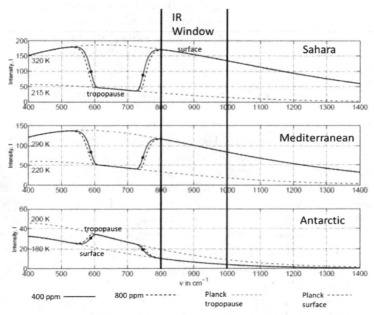

Figure 86. A theoretical, CO2-only model. The solid black line is 400 ppm CO2, dashed is 800 ppm. The figures represent emissions detected by a satellite on a clear, cloud free day.

He then considers the Antarctic temperature inversion, the variation of emission by latitude, and assumes an emissivity of 0.8, rather than one, and determines that the temperature difference between 400 ppm and 800 ppm is 0.92K. This temperature difference accounts for the additional energy retention at 800 ppm. It is similar to other estimates as most show doubling CO$_2$ raises surface temperatures by about one degree. The doubling would raise temperatures in the lower troposphere slightly more, perhaps 1.1 to 1.2°C, due to evaporation, convection, and condensation.

Happer's calculation of the theoretical effect of CO$_2$ on climate is helpful. So far, the man-made greenhouse effect has never been measured or observed in nature. The climate change debate *is* entirely theoretical, yet most have never been through the steps in computing the possible impact of CO$_2$ on climate or global warming. To follow a famous physicist through the process is rewarding.

It also explains where the IPCC and the consensus have gone wrong. They emphasize forcing from the top of the troposphere directed toward the surface. This is backwards, as Happer explains, we need to focus on the path the energy takes from the surface to outer space. CO_2 and water, including ice and water vapor, are important parts of Earth's cooling process. Variations in the rate of cooling, not extra heating, change the surface temperature.

As discussed above, Lindzen shows how nature reduces high-level cirrus clouds to allow more thermal energy to escape when surface temperatures rise. Also, when we look at an example of CO_2 cooling, rather than heating, we see one of the reasons why temperatures in Antarctica are not always consistent with the rest of the planet.

Defining the Greenhouse Effect

As already mentioned, in 2020, William Wijngaarden and Will Happer refined the rough calculations Happer did for the debate and produced a rigorous and precise description of the radiative "greenhouse effect," often abbreviated as "GHE." By way of contrast, the IPCC definition is very ambiguous. In AR6,[171] the IPCC defines the greenhouse effect in several ways. It is the overall warming due to the presence of greenhouse gases in the atmosphere (p. 1-37), it is also the additional warming caused by CO_2 increasing in the atmosphere before or after adding feedback to that warming (pp. 7-65 & 7-66), and it is the difference between incoming radiation from the Sun and outgoing radiation (pp. 7-17 & 7-61). These multiple vague definitions are confusing.

Regardless of the definition, the GHE has never been measured, only modeled. That said, we do know the GHE exists, we can measure the downwelling radiation from the atmosphere from Earth's surface. To make matters worse, the total GHE has numerous possible components and feedbacks, and their relative contributions are unknown.

Basic physics suggests that Earth's surface is warmer than it would be with a transparent atmosphere, that is no greenhouse gases, clouds, or oceans. As discussed above, if we assume Earth is a blackbody, then subtract the solar energy reflected, from the hypothetically non-existent clouds, atmosphere, land, ice, and oceans; we can calculate a surface temperature of 254K or roughly -19°C. The actual average temperature today is about 288.7K or roughly 15.5°C. This modeled difference of 34°C to 35°C is often called the overall greenhouse effect (IPCC, 2021, p. 1-165).

A blackbody absorbs all energy that strikes it, and it emits the same amount of energy, but the wavelength of the emitted radiation is different, and depends upon the blackbody's temperature. Earth is nothing like this. It isn't black and both the atmosphere and the oceans absorb and redistribute solar energy, often the absorbed energy is circulated for a long time, even centuries or

---

[171] (IPCC, 2021)

millennia, before it is reemitted. A blackbody absorbs and reemits energy with a delay of less than a second.

Earth's surface temperature is not constant, like a blackbody's temperature, it varies a lot by latitude, altitude, season, and/or ocean depth. The Moon has a calculated blackbody temperature of 270K, no atmosphere or oceans, and an average temperature, at the equator and low latitudes, of around 215.5K, so even the Moon is not an ideal blackbody.[172]

Some unknown portion of the overall GHE is probably due to GHGs. These include, CO_2 (carbon dioxide), H_2O (water vapor), CH_4 (methane), N_2O (nitrous oxide), and O_3 (ozone). Wijngaarden and Happer examine the likely influence of these greenhouse gases using the HITRAN line-by-line molecular transmission and absorption database maintained at Harvard University (Wijngaarden & Happer, 2020). We will refer to Wijngaarden and Happer's important paper as W&H. HITRAN stands for *high*-resolution *tran*smission molecular absorption. The database compiles spectroscopic parameters that computer programmers can use to model the transmission and emission of light in the atmosphere. W&H use it to model a hypothetical mid-latitude temperature and GHG atmospheric profile and derive representative climate sensitivities to doubling the gases.

Gravity has no direct effect on radiation transfer, but since gravity determines how the air pressure changes with altitude, there is an indirect effect because the air pressure influences the density of greenhouse-gas molecules. Thus, it also has a large effect on their total radiation capture cross sections and how that contributes to the opacity of the lower atmosphere. For cloud-free air, the radiation flux is determined by only two quantities, how the opacity of the atmosphere varies with altitude and how the temperature varies with altitude. W&H took the altitude profiles of both temperature and opacity from experimental observations. The HITRAN data is based on observations. It is not theoretical data.

Because Earth's atmosphere is transparent to most solar radiation, and the Earth's surface is opaque, the surface absorbs twice as much incoming radiation as the atmosphere. Per the laws of thermodynamics, a planet will try to emit as much radiation as it receives. If Earth emits more, it cools; if it emits less, it warms. Earth is never in thermodynamic equilibrium, and Earth's average temperature varies every year from about 12°C in January to 16°C in July.[173] This is significant because the yearly global average temperature has increased less than one degree since the beginning of the 20th century.

The Earth's temperature controls the type of radiation it emits, and it emits mostly in the thermal infrared. The range of emitted frequencies, plotted as wavenumbers, with units of 1/cm, are shown in Figure 87.

---

[172] (Williams, Paige, Greenhagen, & Sefton-Nash, 2017)
[173] (Jones, New, Parker, Martin, & Rigor, 1999)

Both the frequency and the power emitted by GHG molecules are determined by the molecule's kinetic temperature. The y axis in Figure 87 is the spectral flux, or the amount of energy passing through the top of the atmosphere for each frequency on the x axis. In this case frequency is expressed as a wavenumber, or the number of waves per cm.

In Figure 87, the black curve represents the flux with 400 ppm CO_2, today's concentration. The green is with no CO_2 at all, and the red is with CO_2 doubled (800 ppm). The area between the green curve and the black curve represents the forcing due to taking CO_2 from zero to 400 ppm, the tiny area between the black curve and red curve—most of it highlighted in yellow—represents the increase in forcing when CO_2 doubles from 400 ppm to 800 ppm. The point is, CO_2 is spent, there isn't much more it can do beyond what is already done.

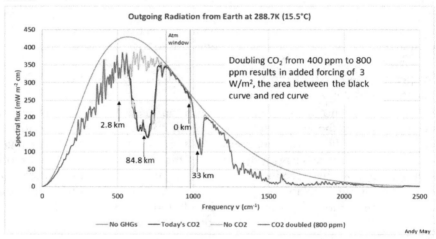

Figure 87. The computed outgoing radiation flux spectrum from Earth, with a temperature of 288.7K. The red, green and black curves are the emitted radiation as modified by CO_2 (in various concentrations) and H_2O, O_3, CH_4, and N_2O, the main GHGs, in their present concentrations. The blue curve is the flux with no GHGs in the atmosphere. The black curve represents the flux with 400 ppm CO_2, today's concentration. The green is with no CO_2 at all, and the red is with CO_2 doubled (800 ppm). The x axis is frequency in wavenumber units and the flux is in (mW cm)/m^2. The additional forcing caused by doubling CO_2, is the integrated area of the distance between the black line and the red line, the major differences are highlighted in yellow. The rest of the figure is explained in the text. The data used is from HITRAN and the figure is very similar to Figure 4 in (Wijngaarden & Happer, 2020).

A perfectly transparent atmosphere would radiate all surface emitted energy according to the blue line in Figure 87. All the curves plotted overlay in the atmospheric window ("Atm window") from 824 to 975 cm^{-1}. In this window the surface radiation can go straight to space, so it is labeled 0 km. This means the radiation at the satellite represents the surface temperature of 288.7K or 15.5°C. Other example departures from the ideal Planck brightness blue curve, are labeled with the approximate altitude of the emissions, based on their brightness temperature. At these frequencies the atmosphere below the emission altitudes can be considered opaque to surface radiation due to the combination of GHGs modeled.

The marked altitude of 84.8 km, in the middle of the CO_2 caused divot in the energy curve, means that the emissions at that frequency, come from that average modeled altitude. Below that altitude, CO_2 blocks the radiation at that frequency. In this critical region of the spectrum, CO_2 is saturated and cannot block more radiation. The emission altitude of 84.8 km is nearly at the top of the atmosphere, very high in the mesosphere (see Figure 79), yet the satellite observes this frequency coming from that altitude. Additional CO_2 makes the divot a little wider though, as the red curve illustrates, and the widening causes a bit more radiation at the edges of the enlarged divot to be blocked.

As CO_2 doubles from today's concentration, the radiation emission level, for each frequency absorbed, moves upward. If the level is in the troposphere and cloud cover does not change, the emission temperature decreases with altitude, so the amount of energy emitted decreases. Since the energy emitted to space is less, Earth's surface warms. When the sky is clear, radiation fluxes can be calculated accurately, but the conversion to a temperature change can get complicated. Emission height is strongly dependent upon frequency, as Figure 87 shows, and an average emission height has little meaning; just as a global average temperature has little practical meaning.

At higher altitudes, in the middle stratosphere (see Figure 79), temperature warms with altitude. The warming is due to an increase in ozone (O_3). This means as the emission level moves higher, more energy is emitted per molecule. This results in cooling, and we see this effect at the bottom of the CO_2 divot in Figure 87. The red and the black curves reverse their positions and added CO_2 causes cooling.

Figure 79 shows an atmospheric temperature profile very similar to the one used for the W&H model. In the real world, the temperature profile varies a lot from location to location and with time, especially in the troposphere, but W&H use a single set of values that are representative of the "standard atmosphere" in the mid-latitudes for their model.

There is very little H_2O above the tropopause and N_2O and CH_4, already minor greenhouse gases in the troposphere, also decrease. In the stratosphere ozone (O_3) warming is dominant and temperature increases, until the mesosphere is reached where ozone decreases rapidly and CO_2 cooling begins to dominate. At the top of the mesosphere, roughly 86 km, changes in the energy flux are negligible, so W&H call this the top of the atmosphere or TOA.

Convection is minimal above the tropopause, but in the troposphere, it provides about half of the total heat transfer from the surface to the tropopause. As Figure 87 shows, except for the atmospheric window, the lower troposphere (below 2.8 km) is mostly opaque to Earth's OLR, even on a cloud-free day. The evaporation of water transports much of the surface emitted thermal energy, as latent heat, to higher altitudes where it can be radiated away from Earth. Thermal energy emissions, released from condensing water vapor, begin to appear at the base of low-level clouds, and continue throughout the cloud. Clouds are a very important component of Earth's cooling system but cannot currently be modeled, so they are not included in the W&H model.

Clouds are included in the IPCC general circulation models (GCMs), but the IPCC assumes the cloud parameters and impact, they cannot calculate them. The IPCC AR6 report acknowledges that "clouds remain the largest contribution to overall uncertainty in climate feedbacks (*high confidence*)." (IPCC, 2021, pp. TS-59). While researchers are trying to model clouds, we share their *high confidence* that clouds are the largest source of uncertainty in computing the impact of humans on climate change.

The W&H emissions model is a clear sky model and only accurate above the clouds and in areas where there are few clouds, such as the poles and over deserts. Clouds excluded, W&H do try and account for GHG warming feedbacks. They take into account the vertical distribution of the GHGs, for example water vapor virtually disappears above the troposphere, but CO_2 keeps the same concentration until high in the mesosphere.

They investigated three cases, fixed relative humidity with a constant tropospheric lapse rate, fixed relative humidity with a variable lapse rate, and fixed absolute humidity. The latter case is the least likely, and it results in 1.4°C of warming when CO_2 reaches 800 ppm. Fixed relative humidity with a variable lapse rate results in 2.2°C, and a constant lapse rate of 6.5 km^{-1}, results in 2.3°C of warming. None of these scenarios are alarming.

W&H compared their model to actual satellite measurements from the Antarctic, Sahara Desert, and the Mediterranean region. All comparisons were very good. The curves they calculated were like those shown in Figure 85.

In summary, W&H have provided us with a detailed and accurate emissions model that shows only modest warming (2.2 to 2.3°C), inclusive of likely water vapor feedback, but not counting the feedback due to changes in cloudiness. Both the magnitude and sign of net cloud feedback to surface warming are unknown. Lindzen has shown it is likely negative (cooling) in the tropics, but outside the tropics no one knows.

They computed the change in forcing for a doubling of each GHG. Doubling the methane or nitrous oxide concentration changes the outgoing forcing by less than one percent, thus they are not significant. Due to the properties of water vapor, its atmospheric concentration is very unlikely to double, but if it did, it would only increase the forcing by eight percent at 11 km. Doubling CO_2 only increases the forcing by four percent at 11 km. In the mid-latitudes, 11 km is the top of the troposphere.

Doubling CO_2 will cause the stratosphere to cool about 10°C, but the changes in surface temperatures from this model are warming of 2.2 to 2.3°C. This is much less than the preferred IPCC AR6 value of 3°C (IPCC, 2021, pp. TS-57). Considering that the current net effect of clouds is cooling, and it seems likely that total water vapor in the atmosphere is decreasing or staying flat, these results suggest we have little to worry about regarding increasing GHGs.

Convection

As Happer mentions in his Major Statement, more solar radiation is absorbed in the tropics than at the poles. Due to greenhouse gases and clouds, much of this absorbed energy cannot be radiated directly to space, so it needs to move to an altitude or a location where it can be radiated away.

Background: General Atmospheric Circulation

Convection plays a large role in cooling the troposphere. The lower troposphere (boundary layer) is opaque to most of the outgoing IR radiation emitted by Earth's surface and only allows about 10 to 30% to be emitted directly to outer space. Most of this is through the IR window. Most of the greenhouse effect and opacity of the boundary layer is due to water vapor, which is unevenly distributed in the troposphere. The poles and deserts have little, the tropics have a lot. Water vapor is less dense than dry air, which causes it to rise until it condenses into clouds, rain, or snow. The combination of the uneven distribution and low density of water vapor contribute to the air circulation patterns shown in Figure 88, which is from Happer's Major Statement.

The Role of the Tropics in the General Circulation

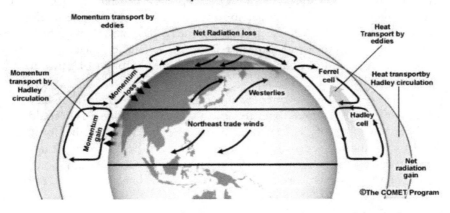

Figure 88. Air circulation patterns and net surface radiation to space. Tropical heat loss speed, due mainly to evaporation, drives the circulation engine.

By far, most sunlight is absorbed by the tropics, where the Sun is nearly directly overhead. The hot Sun causes an increase in vapor pressure over the ocean, and surface water in general, increasing evaporation and equatorial tropospheric water vapor. Water vapor is the most powerful GHG. Clouds are widespread in the tropics, and in combination with the humid air, make the tropics almost entirely opaque to much of the surface emitted IR.

The tropics lose this accumulated heat via convection. The Earth spins from west to east, spinning fastest in the tropics. Tropical circulation is in the opposite direction and the wind moves from east to west. As Figure 88 shows, net radiation gain is greatest in the tropics and the region adds a lot of energy to the global circulation pattern, and it gains momentum.

As dictated by the second law of thermodynamics, energy will flow from warmer areas to cooler areas. So, the net flow is away from the equator and toward the poles. There is a net absorption of solar energy and a positive energy balance until some point in the mid-latitudes, where the net gain of energy turns into a net loss and the balance becomes increasingly negative moving poleward. The change in balance is indicated by a change from red to blue shading in Figure 88. The North Pole is a net emitter of radiation. Some of the energy is lost to the surface, warming it, and some is radiated upward to space—if it can escape being trapped by water vapor. Radiation exiting to space from the upper troposphere is derived from both water vapor and CO_2. The ratio of CO_2 to water vapor emissions increases with altitude. CO_2 is a critical cooling agent for the upper atmosphere.

Background: Water vapor

Tamblyn claims that as atmospheric temperature rises, so does total atmospheric water vapor. Earlier in the chapter we saw that this may not be true in nature (see Figure 75). As water vapor is a powerful GHG, if CO_2 causes temperature to rise and temperature causes water vapor to rise, the temperature increase will accelerate, a positive feedback. He supports this claim with a reference to a paper by Brian Soden and colleagues, who base their conclusion on a model of the 1991 volcanic eruption of Mt. Pinatubo (Soden, et al. 2002).

Soden and colleagues acknowledge that, while their Pinatubo model is consistent with a positive water feedback, changes in cloud cover could have had the same effect. Volcanic eruptions produce aerosols that provide a nucleus for the water droplets that form clouds. Depending upon the nature and power of the eruption, and Mt. Pinatubo was a huge eruption, these volcanic aerosols can cause many changes in clouds at all altitudes, and over very wide areas.

As we have seen, clouds are a huge unknown in these model experiments, because their impact can be large. Soden assumed that the impact of clouds did not change after the eruption or during the recovery from the volcanic cooling. Thus, while he showed positive feedback of some sort is needed to model the volcanic eruption and its effects on climate, the idea that the feedback came from changes in atmospheric water vapor content was an assumption. Changes in cloud cover, due to the presence of volcanic aerosols, could obviously influence climate and water vapor concentration. As we have seen, cloud formation and destruction cannot currently be modeled (Nakamura, 2018). Mototaka Nakamura, a well-known expert in climate models, writes:

> "The vertical diffusion of water vapor, however, has very complicated effects on the radiative forcing profile of an atmospheric column and [it] is not clear to me if it is contributing to artificial warming at the surface. … A profound fact is that only a very small change, so small that it cannot be measured accurately with the currently available observational devices, in the global cloud characteristics can completely offset the warming effect of the doubled atmospheric carbon dioxide." (Nakamura, 2018)

The Mt. Pinatubo eruption was large enough to cool the entire planet. At the same time, the global water-vapor concentration dropped. Soden's work was done by modeling the decline in temperature after the eruption and then modeling the later warming. He successfully compared his modeled temperatures and water-vapor concentrations to observations at the surface and higher in the atmosphere. Is this definitive evidence that water-vapor concentration is solely controlled by surface temperature? Probably not, it supports the idea, but that is all.

Other data on the relationship between atmospheric temperature and total water-vapor content is more ambiguous and various datasets do not agree with one another.[174] Logically, we would expect atmospheric water vapor to increase as temperatures go up, higher temperatures, *all other factors held constant*, should lead to more evaporation and more atmospheric water vapor. But the cloud response to both water-vapor concentrations and temperature is unclear, and the data supporting the direct temperature-to-water-vapor connection is weak. As mentioned at the beginning of this chapter, some datasets show atmospheric water vapor increasing while others show it decreasing. There are too many unknowns to take Soden's speculation seriously.

We discussed upper-troposphere-tropical predicted versus actual temperatures in Chapter 5 and showed that an accurate climate-model prediction has not been observed in the real world. With all the recent focus on CO_2, the importance of water vapor and clouds is often ignored. This distorts our perception of the uncertainty in the climate data. Small changes in either the cloud coverage or properties or in atmospheric water vapor can completely swamp the CO_2 effect.

---

[174] (Benestad 2016) (Miskolczi 2010)

Climate models show an increase in atmospheric water vapor above four kilometers with increasing temperature, but all levels above the top of the boundary layer, show a decline in the tropics and the southern mid-latitudes in the NCEP weather reanalysis dataset from 1973 to 2007.[175] Garth Paltridge and colleagues found an increase in water vapor below 850 mbar (about 1.5 km) consistent with rising temperatures, but a decrease above that. This suggests that there are other factors, besides temperature, that affect total atmospheric water vapor. Cloud formation may well be one of them.

Tamblyn's assertion, that water vapor increases with temperature, is a widely held belief and often heard. It is logical, but not seen in nature—at least not consistently. The scientific approach to this sort of problem is to acknowledge the problem and look for a solution.

Paltridge and colleagues acknowledge that there are potential problems with the weather-balloon instruments they used in their study, but they do not think their results should be ignored or "written off" simply on the basis that they contradict the climate models. Unfortunately, all too often, we see a denial that the data are correct and an insistence that the models are correct.

Numerical Modeling

Climate modelers have a tough job, as Happer describes in his Major Statement. On pages 23 through 25, Happer describes the five equations and numerous variables needed to describe the properties of a parcel of air in the atmosphere. The general-circulation models that the IPCC relies upon to predict future climate use variations of these equations to model fluid and heat flow in the atmosphere and upper ocean.

They model the state of the climate through many numerical quantities, especially wind velocity of "air parcels" at numerous positions on Earth's surface. All the air in a single parcel has the same velocity, temperature, and pressure. The volume of Earth's atmosphere is roughly 5×10^{10} cubic kilometers. Happer writes that if each air parcel were to be one cubic kilometer, we would need 255 billion numbers to characterize the dry state of the entire atmosphere.

The laws of fluid flow include the Navier-Stokes equation, which describes the movement of a compressible fluid such as air. We also need equations to describe the interaction of radiation with mass analogous to the Planck and the Schwarzschild equations. Finally, we need to describe the amount of energy in a parcel that is unavailable to do work. This is often called entropy. Classical theory suggests maximum entropy is achieved when an air parcel is at thermal equilibrium and neither warming nor cooling, but this concept is hotly debated.[176] Maximal entropy is achieved by the thermodynamic tendency

---

[175] (Paltridge, Arking, & Pook, 2009)
[176] (Verkley & Gerkema, 2004)

for energy to flow from warmer to cooler objects. With air parcels, the problem gets very complex, very quickly because they move and their movement prevents thermal stability, a requirement for each climate-model grid cell.

With these equations we can describe the properties of a parcel of dry air. Water vapor and clouds cannot be modeled this way. They involve microscopic processes, but in aggregate, they dominate our climate. Because they cannot currently be modeled, the modelers must parameterize them—meaning they guess.

Pollution

Happer enthusiastically supports fighting real pollution, like ash, oxides of sulfur and nitrogen, or water pollution. He also lists carbon monoxide and soot from indoor stoves as serious pollutants in the third world. These are both carbon-based, but deadly. In the western world we do not normally cook or heat our homes with wood or dung, which is a good thing, indoor air pollution from doing so causes as many as 1.6 million deaths per year.[177] The World Health Organization estimates the figure to be four million deaths per year. A discussion of both figures can be read at OurWorldinData.org. CO_2 is also carbon-based, but essential for life on Earth. He firmly believes CO_2 is not a pollutant and writes:

> "I regard the war on the responsible use of coal and other fossil fuels as deeply immoral. It will impoverish most people by raising the cost of energy. … When used responsibly, fossil fuels release negligible real pollutants like fly ash, oxides of sulfur, nitrogen, etc. The much-demonized CO_2 that must be released, along with H_2O, is actually a benefit to the earth, not a pollutant." (Happer, 2021a, p. 50)

Karoly will not call CO_2 a pollutant, but he firmly believes that CO_2 is causing warming and that the warming is dangerous. He writes:

> "Hence, any emissions of carbon dioxide from human activity contribute to global warming, irrespective of when they occur. Any delay in reducing emissions associated with continued burning of fossil fuels increases global warming as it adds to the level at which global average temperature will eventually stabilize. Many studies have shown that most of the existing fossil fuel reserves must remain unburned, and in the ground if we are to limit global warming to below two degrees." (Karoly, 2021a, p. 32).

---

[177] Institute for Health Metrics at the University of Washington

Tamblyn has no problem calling CO_2 a pollutant. He refers to U.S. law, where the definition of a pollutant is quite broad. Air pollution is any substance that is emitted into the atmosphere, or any substance that results in those emissions, that damages the welfare of humans, animals, or property. He notes that thermal pollution is a well-established principle in water management. So, he pulls all this together and, since he believes that global warming is damaging humanity and CO_2 causes it, CO_2 must be a pollutant (Tamblyn, 2021a, p. 18). He believes that Happer's comments regarding the topic are irrelevant. It seems Tamblyn believes the term pollutant can be used even when there is no evidence a particular chemical or gas causes any harm to human health or the environment.

We have covered many details of the arguments made on both sides in this debate. What details do the participants believe are the most important for their side of the debate? What details are the weakest for the other side of the debate? We look at their picks next.

Chapter 17: The Strongest and Weakest arguments for both sides?

> "It is better to debate a question without settling it than to settle a question without debating it." Joseph Joubert, French philosopher.

At the end of their interviews both Karoly and Happer were asked to list what they consider the five strongest arguments supporting their view of climate change and the five weakest arguments from their opponents. Their choices were illuminating.

The Strongest Arguments for my side

Karoly's first argument was that increasing greenhouse-gas concentrations have been primarily due to human activities. It is a fair point and not disputed by Happer.

Happer's first point was that the climate models have predicted much more warming than has been observed. He also noted that this is strong evidence that the climate sensitivity is much less than the three degrees claimed by the IPCC. It is a point we discussed in detail in Chapters 5, 9, and 16. The consensus has been trying unsuccessfully to prove that humans are causing dangerous climate change for over thirty years, during this time they have always predicted more warming than ultimately occurred. At some point enough is enough. In Aesop's fable of the boy who cried wolf, the boy was ignored after twice crying "wolf." The IPCC have issued six major reports that predicted doom over the past 30 years, none have come true, at some point we must ignore them. If a real climate problem "wolf" emerges, we will deal with it. Repeatedly warning us about nonexistent problems, does not help us, it hurts us.

Karoly's second point was that the observed large-scale increase in global surface temperature has been due to human activity. This is an assertion and opinion. Clearly, neither he, Tamblyn, nor the IPCC have made an effective case. As we saw in Chapter 9, humans probably have some impact on the climate; many species, such as trees and grass, do also. But the amount of human impact is unknown.

Happer's second point was that the consensus has largely ignored the benefits of CO_2. It is almost certainly true and was discussed in detail in Chapter 10. There is a tendency by the consensus to simply assume global warming and more CO_2 are bad. Their arguments tend to refer to a perfectly ordinary event, such as a tornado or hurricane and say, "See, that is the result of global warming." Every bit of extreme weather becomes proof of anthropogenic climate change. As Karl Popper explained, this is a sure sign of pseudoscience (Popper, 1962).

Karoly's third point was also an opinion without sufficient evidence. He stated there would continue to be significant global warming over the twenty-first century, with its magnitude dependent on the emission of greenhouse gases from human activity. It is an opinion obviously based upon modeled projections of Earth's climate and would carry some weight if those projections proved to be accurate. Unfortunately for Karoly, as we saw in Chapter 5, they have not.

Happer's third point was that the Medieval Warm Period and the Little Ice Age occurred before the widespread use of fossil fuels—a point that could be extended to the Paleocene-Eocene Thermal Maximum. Chapters 2 and 3 cover these topics.

Karoly's fourth point was that there have been substantial adverse impacts on human and natural systems from global warming. We dealt with these issues in Chapters 8 and 10. There undoubtedly have been some negative impacts in some places due to climate changes. There always have been and always will be. Climate is a regional thing, and any change will affect different places in different ways, thus some areas will benefit from the changes and others will not. But globally, warming and additional CO_2 have benefited humanity. The world is now greener and agricultural productivity is up. Climate is something best dealt with locally, attempting to deal with it globally will benefit very few people, and probably harm most, as discussed in Chapter 13.

Happer's fourth point was that there is a strong correlation of temperature with solar activity. This was discussed in detail in Chapter 4. Karoly and the IPCC tried to ignore evidence that the Sun contributed to the late-twentieth century warming. Avoiding valid evidence that solar variability affects our climate weakens their case.

Karoly's fifth point was that rapid and sustained reductions in greenhouse-gas emissions from human activities are needed to slow global warming and stabilize global temperature at a safe level. This topic was covered in Chapters 12 and 13.

Happer's fifth point was that the frenzied ad hominem attacks on credible opponents in the climate debate show that the consensus opinion on climate change is weak. He continued that you don't need ad hominem attacks if you have strong scientific arguments. Formal debates do not allow personal attacks, they are based upon reason and evidence, this topic was covered in Chapter 1.

Karoly's first argument that increasing GHGs in the atmosphere are due to human emissions was fine. Here, the opponents agreed. His second through fifth arguments were opinions with little foundation. In debates, assertions require foundation, and he provided none, other than references to the IPCC reports. The IPCC model results and projections he relied on can be discounted. The models are clearly not accurate or predictive, as shown in Chapter 5.

Karoly's best argument in the debate, in our opinion, was the atmospheric fingerprint discussed in Chapter 4. It was related to his point that, if the Sun were truly the major climate driver, the stratosphere should be warming, not cooling. The fingerprint was circumstantial evidence, and factors other than human activities, such as decreasing ozone, could be causing the stratosphere to cool. Further, while the well-mixed gas, CO_2, might be cooling the stratosphere, where water vapor is absent, it does not mean CO_2 plays a major role in tropospheric warming. Both the cooling and warming mechanisms in the troposphere are different from those in the stratosphere, as discussed in Chapters 4 and 16. But, regardless of the problems with the fingerprint hypothesis, it was still his strongest argument. He didn't put it in his list, but we felt compelled to add it.

The Weakest Arguments for the other side

Karoly believed the weakest argument from the skeptical side was that the observed warming over the past century has been due to natural variability. He does not think that solar variations or other natural causes, as we emerge from the Little Ice Age, have been strong enough to cause warming to such a degree. These ideas were discussed in Chapters 2 through 4.

Happer thought the weakest argument from the consensus was the idea that 97% of scientists agree with the consensus (discussed in Chapter 11). Science has absolutely nothing to do with consensus, science is all about reproducible results. Consensus is all about opinions and politics.

Karoly's second weak argument from the skeptics was that observed increases in greenhouse gases have been from natural sources. Happer agreed with this point, probably most of the increase in atmospheric CO_2 has been from human emissions, with the remainder outgassed from the oceans due to increasing temperatures.

Happer's second weak point from the consensus was, "temperature has increased for the past century and CO_2 levels have increased. Therefore, the temperature increase was caused by CO_2." It has been the main argument the consensus has used, and it is a logical fallacy. Correlation is not causation.

Karoly's third weak point was, "Even if the carbon dioxide increases are due to human activity, increases in carbon dioxide are good for plants and agriculture." It was very odd that Karoly would choose this point to highlight, since the evidence of the agricultural benefits of additional CO_2 are so obvious and well documented. Recall what we learned in Chapter 10.

Happer's third weak point was, "Increasing CO_2." What he was getting at is the tendency, on the part of the consensus, politicians, and the media, to assume additional CO_2 is bad. It is literally the basis of all life on Earth. All the oxygen we breathe came from CO_2, through the miracle of photosynthesis.

We owe our existence to CO_2, in and of itself, we know it is beneficial. This is discussed in Chapter 10.

Karoly's fourth weak skeptical point was, "Even if the increases of greenhouse gases in the atmosphere cause some global warming, this warming will be mainly beneficial." Economically this clearly has been true, as described by economists Richard Tol and William Nordhaus. Nordhaus won the Nobel Prize for his work on the economics of climate change. We covered their analyses in Chapters 12 and 13. The benefits of warming and additional CO_2 have been very apparent today, as described in Chapter 10.

Happer's fourth weak consensus point was, "Government-funded, consensus-supporting researchers have no conflict of interest." Obviously, government funding involves a conflict of interest. There has always been an agenda behind government funding, and it is generally political as opposed to scientific (recall Chapters 12 through 15). Unfortunately, due to typical government over-reaction, a large climate change research industry has been built around the world dependent upon government and U.N. funding. This industry is supported by billions of taxpayer dollars and will not go quietly. As President Ronald Reagan pointed out on August 8, 1992, "Nothing lasts longer than a temporary government program." Thus, the climate-industrial complex keeps beating the dead horse, hoping, against hope, that it will revive. Six IPCC reports attest to their persistence.

Karoly's final weak skeptical point was simply a re-statement of his fourth. He wrote, "Even if there are some adverse impacts from human-caused global warming, it will cost too much to reduce GHG emissions and it will be cheaper for modern society to adapt to the impacts of climate change rather than to reduce emissions from burning fossil fuels." Many do not believe this is a weak argument and believe it to be true, including William Nordhaus and William Happer. Nordhaus showed that reducing GHG emissions as rapidly as the IPCC recommends will cost the world $50 trillion out of an $88 trillion economy.

Happer's last consensus weak point was, "Scientific opponents of the consensus are prostitutes of the evil fossil fuel industry" (Happer, 2021a, p. 55). This statement, or something similar, is often made by the consensus. It is wrong on so many levels. The fossil fuel industry has funded very little climate research, certainly much less than the billions of dollars of funding from governments intended to prove humans cause climate change. Examples of government efforts to dictate climate research outcomes abound, many were discussed in Chapters 14 and 15, as well as in our earlier book, *Politics and Climate Change: A History*. But, even if a fossil fuel company did fund climate research, why is that wrong? Fossil fuel companies are free to fund research, just as governments are. Regardless of governments and the consensus trying to change observations to match their political agenda, facts are stubborn things. How many times can the governments and the consensus cry wolf?

Chapter 18: Final Thoughts

"The whole point of science is to question accepted dogmas." Freeman Dyson to the *Boston Globe*, November 5, 2013. (Heartland Institute, 2015)

Karoly, Tamblyn, and Happer agree that the contribution of human emissions to modern warming cannot be measured and must be computed with climate models. Karoly and Tamblyn think the IPCC models are accurate and use them to predict dangers to mankind and the planet. Happer disagrees and provides evidence the models are not accurate and compute too much warming. He also provides us with an elegant and novel calculation of the direct impact of CO_2 on temperature of about one degree Celsius per doubling of CO_2 (the CS). If this value is correct, it will take 320 years for temperatures to rise two degrees over pre-industrial (Little Ice Age) levels.

Climate sensitivity to CO₂ is key

The debate clarifies a key point. The alleged dangers due to man-made climate change hinge on CS. Is it about three degrees Celsius as Karoly and Tamblyn believe or is it more like one as Happer believes? Is warming of two degrees a hazard or will we be fine up to higher temperatures? Even the assertion that two degrees above the end of the Little Ice Age is a dangerous temperature rise is speculative and controversial. The current global average temperature is about 15°C, but over the last 500 million years the overall average surface temperature has been estimated to be 17 to 20°C, two to five degrees higher than today.[178]

Karoly and Happer have significant and substantive disagreements regarding what temperature is dangerous. All the projected future hazards to mankind and the planet hinge on the value of CS and on the global average temperature considered to be dangerous. While they disagree on the probable value of CS, they agree it is a key unknown and the crux of the debate.

In a January 31, 2021, email to the author, Karoly writes, "A key question for projections of future climate change is the uncertainty in global equilibrium climate sensitivity to increases in carbon dioxide." Karoly then refers to a paper on climate sensitivity by Steven Sherwood and colleagues.[179] The paper uses statistical techniques to assess the range of likely climate-sensitivity estimates.[180] It also acknowledges the accepted range of values in climate sensitivity (1.5°–4.5°C) has not been narrowed in 40 years. AR6 provides us with a range of 2° – 5°C, but the uncertainty is still three degrees. The authors make a case that they have narrowed it to 2.6° to 3.9°. This is a 50% reduction

---

[178] (Scotese, 2015) and (Scotese, Song, Mills, & Meer, 2021)
[179] Climate Change Research Centre at the University of New South Wales
[180] (Sherwood, et al., 2020)

in uncertainty, but their lower bound of 2.6°C/2xCO2, is much higher than many modern observation-based estimates.

There are many problems with Sherwood and colleagues' analysis. They do not consider that net cloud feedback may be negative, when CERES data suggest it is. They discount Lindzen's iris effect and conclude it is not significant, regardless of the climate-model results supporting his observation-based studies. They claim "global-climate models ... are not trustworthy" for evaluating the iris effect, yet they trust them over observation-based studies to compute CS. The paper is full of contradictions.

As Lindzen, Sherwood and colleagues, and many others have concluded, cloud cover feedback is, by far, the largest component in CS uncertainty. Change in cloud-cover is key to determining both the magnitude and the sign (positive or negative) of feedbacks to CO_2-caused warming. Yet, cloud feedback is unknown, and its possible values cover a huge range.

Sherwood and colleagues give the long-term range of cloud feedback as zero to 1.2 W/m². This is supported by studies including one by Ian Williams and Raymond Pierrehumbert[181] and others. Williams and Pierrehumbert found that errors in satellite measurements of incoming and outgoing radiation were large; and that determining the net inflow or outflow of radiation from Earth was difficult using currently available technology. While true, this does not mean their model is correct, only that the uncertainty is high.

Cloud-cover feedback is a complex topic and will not be resolved soon. In general, the Sherwood study emphasizes model results (while claiming they cannot be used to verify the iris effect) and mostly ignores direct observations. The range of time through which we have good observations of clouds from space is far too short and the records are too inaccurate to be definitive.

Literature review papers, like Sherwood and colleagues', are not helpful when they pick and choose what they find as "trustworthy." There is a real need to identify the actual climate sensitivity to additional CO_2, but the value will not be determined by cherry-picking studies. Trying to narrow the range from 1.5° to 4.5° down to 2.6° to 3.9°, when well supported studies—based on observations—provide credible, and unrefuted values below 1.5° is disingenuous. Until better data are available, the best answer to what the impact of CO_2 and clouds are on climate is: "We don't know."

Currently CO₂ is beneficial

Currently, as Happer points out, global warming and the additional CO_2 are beneficial, and will be for many more years. The "dangers" rely on an assumed and unsubstantiated "tipping point" of two degrees Celsius above

---

[181] (Williams & Pierrehumbert, 2017)

either today's, or the 19th century's, average temperature, depending upon who is speaking.

Thus, it appears that Karoly's statements that human-caused climate change is "virtually certain" to be dangerous and we must use "immediate, stringent measures" to deal with it, are unproven.

Happer's initial statement was, in part:

> "There is no scientific basis for the claim that increases in atmospheric CO_2 due to burning of fossil fuels will cause climate change that will have substantial adverse impacts on humanity and natural systems."

This can be considered true. All the proposed hazards of CO_2 hinge on projected warming using a poorly estimated value of climate sensitivity. Low values of CS are no problem. High values may be a problem, but no one knows the correct value or even the dangerous value.

The last part of Happer's initial statement is:

> "Any resulting climate change will be moderate, and there will be very major benefits to agriculture and other plant life."

Because the CS is so poorly understood, the first part of his statement is stated too firmly. It is safe to say any human-caused climate change will *probably* be moderate. The last part of the statement is fine, we can be certain that there will be major benefits to agriculture and other plant life, that is clearly happening already.

Generally, the debate was informative, educational, and interesting. The material will not be new to many readers, but Karoly and Happer are well-known, credible experts and that gives extra weight to their remarks and opinions. Thankfully, they did debate the topic! The consensus has avoided debates, presumably to keep from educating the public. Once the true scientific underpinnings of climate alarmism are known, the public will see how fragile and delicate they are.

According to the U.S. Government Accountability Office, the U.S. government alone spent $154 billion[182] through 2017 on climate science and related activities. It is notable that, after all this spending and time, the estimated range of climate sensitivity to CO_2 is unimproved and the uncertainty is still 3°C. We are no closer to determining what level of warming is dangerous or harmful. These two key questions remain unanswered.

---

[182] (Government Accounting Office, 2018)

Tamblyn's arguments are weaker than those of the other two. They are mostly conjecture, using unstated and unrealistic assumed temperature increases and impacts with little foundation. Happer describes them well in his final reply:

> "Mr. Tamblyn has produced not so much a response to my Statement, as a primer on global-warming alarmism, a whole list of scary talking points and computer-generated graphs, with occasional asides to deplore how obtuse I am for not understanding the gravity of this supposedly existential threat to the planet or how ignorant I am of basic physics."

That last line is a bit absurd; Professor Happer is one of most accomplished and brilliant physicists alive today. Tamblyn did not pick up where Karoly left off. He chose to begin a new debate by listing a lot of very speculative claims about the dangers of warming. Most of these dangers involved warming rates and amounts that are completely unrealistic. Unfortunately for us, what he adds is mostly irrelevant to the original discussion between Happer and Karoly. Thus, Happer had to defend his views in two separate debates, but he soldiered on and did quite well in both. It is a shame Karoly backed out in the middle of the debate. Had he stayed, the result would have been better. As it is, we at least gained something of great value: we found out it is all about how much and how fast our Earth is warming. In my opinion, Happer won the debate: he won them both.

Works Cited

Archer, D., & Ganopolski, A. (2005). A movable trigger: Fossil fuel CO2 and the onset of the next glaciation. *6*(5). Retrieved from https://agupubs.onlinelibrary.wiley.com/doi/full/10.1029/2004GC000891

Baliunas, S., Donahue, R., Soon, W., Horne, J., Frazer, J., Woodard-Eklund, L., . . . Rao, L. (1995). Chromospheric variations in main-sequence stars. *The Astrophysical Journal, 438*, 269-287. Retrieved from http://adsabs.harvard.edu/full/1995ApJ...438..269B7

Ball, T. (2014). *The Deliberate Corruption of Climate Science.* Seattle: Stairway Press. Retrieved from https://www.amazon.com/Deliberate-Corruption-Climate-Science-ebook/dp/B00HXO9XGS/ref=tmm_kin_swatch_0?_encoding=UTF8&qid=&sr=

Barnett, T. P., Pierce, D. W., Achutarao, K. M., Gleckler, P. J., Santer, B. D., Gregory, J. M., & Washington, W. M. (2005, July 8). Penetration of Human-Induced Warming into the World's Oceans. *Science, 309*(5732), 284-287. Retrieved from https://science.sciencemag.org/content/309/5732/284.abstract

Beerling, D., & Royer, D. (2011). Convergent Cenozoic CO2 history. *Nature Geoscience, 4*. Retrieved from https://www.nature.com/articles/ngeo1186

Behringer, W. (2010). *A Cultural History of Climate.* Cambridge, UK: Polity Press. Retrieved from https://www.amazon.com/Cultural-History-Climate-Wolfgang-Behringer/dp/0745645291

Benestad, R. (2016, January 21). A Mental Picture of the Greenhouse Effect. *Theoretical and Applied Climatology, 128*(3-4), 679-688. Retrieved from https://link.springer.com/article/10.1007/s00704-016-1732-y

Berner, R., & Kothavala, Z. (2001). Geocarb III: A Revised Model of Atmospheric CO2 over Phanerozoic Time. *American Journal of Science, 301*(2), 182-204. Retrieved from http://www.ajsonline.org/content/301/2/182.short

Bindoff, N., & Stott, P. (2013). *Detection and Attribution of Climate Change: from Global to Regional Supplementary Material.* IPCC. Retrieved from https://www.ipcc.ch/site/assets/uploads/2018/07/WGI_AR5_Chap.10_SM.pdf

Black, B., Neely, R., Lamarque, J.-F., Elkins-Tanton, L., Kiehl, J., & Shields, C. (2018, November 30). Systemic swings in end-Permian climate from Siberian Traps carbon and sulfur outgassing. *Nature geoscience, 11*, 949-954. Retrieved from https://www.nature.com/articles/s41561-018-0261-y

Blair, T. (2013, January 17). Children just aren't going to know what snow is. *Daily Telegraph.* Retrieved from https://www.dailytelegraph.com.au/blogs/tim-blair/children-just-arent-going-to-know-what-snow-is/news-story/5a16c85680b7cc94f345240a727fb09d

Blunden, J., & Arndt, D. S. (2020). *State of the Climate in 2019.* BAMS. Retrieved from https://www.ametsoc.org/index.cfm/ams/publications/bulletin-of-the-american-meteorological-society-bams/state-of-the-climate/

Buis, A. (2020, February 27). *Milankovitch (Orbital) Cycles and Their Role in Earth's Climate.* Retrieved from NASA:

https://climate.nasa.gov/news/2948/milankovitch-orbital-cycles-and-their-role-in-earths-climate/

Burgess, M., Ritchie, J., Shapland, J., & Pielke Jr., R. (2020, December 23). IPCC baseline scenarios have over-projected CO2 emissions and economic growth. *Environmental Research Letters*. Retrieved from https://iopscience.iop.org/article/10.1088/1748-9326/abcdd2

Burgess, S., Bowring, S., & Shen, S.-z. (2014, March 4). High-precision timeline for Earth's most severe extinction. *PNAS*, 3316-3321. Retrieved from https://www.pnas.org/content/111/9/3316

Ceppi, P., Brient, F., Zelinka, M., & Hartmann, D. (2017, July). Cloud feedback mechanisms and their representation in global climate models. *WIRES Climater Change, 8*(4). Retrieved from https://onlinelibrary.wiley.com/doi/full/10.1002/wcc.465

Charney, J., Arakawa, A., Baker, D., Bolin, B., Dickinson, R., Goody, R., . . . Wunsch, C. (1979). *Carbon Dioxide and Climate: A Scientific Assessment*. National Research Council. Washington DC: National Academy of Sciences. Retrieved from http://www.ecd.bnl.gov/steve/charney_report1979.pdf

Christiansen, B., & Ljungqvist, F. C. (2012). The extra-tropical Northern Hemisphere temperature in the last two millennia: reconstructions of low-frequency variability. *Climate of the Past, 8*, 765-786. doi:10.5194/cp-8-765-2012

Christy, J. (2019, June 18). *Putting Climate Change Claims to the Test*. Retrieved from Global Warming Policy Forum: https://www.thegwpf.com/putting-climate-change-claims-to-the-test/

Christy, J., Spencer, R., & Braswell, W. (2000, September). MSU Tropospheric Temperatures: Dataset Construction and Radiosonde Comparisons. *Journal of Atmospheric and Oceanic Technology, 17*, 1153-1170. Retrieved from https://journals.ametsoc.org/view/journals/atot/17/9/1520-0426_2000_017_1153_mttdca_2_0_co_2.xml

Clark, T., Raby, G., & Roche, D. (2020). Ocean acidification does not impair the behaviour of coral reef fishes. *Nature, 577*, 370-375. Retrieved from https://www.nature.com/articles/s41586-019-1903-y

Clintel.org. (2021, January 27). *Press release: adaptation much more profitable than mitigation.* Retrieved from https://clintel.org/press-release-adaptation-much-more-profitable-than-mitigation/

Coakley, J. A. (2003). Reflectance and Albedo, Surface. *Encyclopedia of Atmospheric Sciences*. Retrieved from 10.1016/B0-12-227090-8/00069-5.

Connolly et al., R. (2021). How much has the Sun influenced Northern Hemisphere temperature trends? *Research in Astronomy and Astrophysics, 21*(6). Retrieved from http://www.raa-journal.org/raa/index.php/raa/article/view/4920/6080

Connolly, R. (2019, September 23). *How the UN's climate change panel created a "scientific consensus" on global warming*. Retrieved from Medium: https://medium.com/@ronanconnolly/how-the-uns-climate-change-panel-created-a-scientific-consensus-on-global-warming-a062f5f54ab2

Works Cited

Cook, J. (2010, July 30). *10 Indicators of a Human Fingerprint on Climate Change.* Retrieved from skepticalscience.com: https://skepticalscience.com/10-Indicators-of-a-Human-Fingerprint-on-Climate-Change.html

Cook, J. (2011, July 27). *How we know we're causing global warming in a single graphic.* Retrieved from skepticalscience.com: https://skepticalscience.com/How-we-know-were-causing-global-warming-in-single-graphic.html

Cronin, T. (1982). Rapid Sea Level and Climate Change: Evidence from Continental and Island Margins. *Quaternary Science Reviews, 1*(3). Retrieved from https://www.sciencedirect.com/science/article/abs/pii/0277379182900099

Daly, J. L. (1997, June). *A Discernible Human Influence.* Retrieved from Still Waiting for Greenhouse: http://www.john-daly.com/sonde.htm

Dippery, J., Tissue, D., & Thomas, R. (1995). Effects of low and elevated CO_2 on C3 and C4 annuals. *Oecologia, 101*, 13-20. Retrieved from https://link.springer.com/article/10.1007/BF00328894

Donohue, R., Roderick, M., McVicar, T., & Farquhar, G. (2013, May 15). Impact of CO_2 fertilization on maximum foliage cover across the globe's warm, arid environments. Retrieved from https://agupubs.onlinelibrary.wiley.com/doi/full/10.1002/grl.50563

Eddy, J. (1976). The Maunder Minimum. *Science, 192*(4245). Retrieved from http://www.odlt.org/dcd/docs/john_eddy_Maunder_Minimum.pdf

Eisenhower, D. D. (1961, January 17). *Farewell Address to the Nation.* Retrieved from mcadams.posc.edu: http://mcadams.posc.mu.edu/ike.htm

Enserink, M. (2021). Does ocean acidification alter fish behavior? Fraud allegations create a sea of doubt. *Science.* Retrieved from https://www.sciencemag.org/news/2021/05/does-ocean-acidification-alter-fish-behavior-fraud-allegations-create-sea-doubt

ERK. (2010). *The Planck Function.* Retrieved from http://www.atmo.arizona.edu/students/courselinks/fall10/atmo551a/Planck_Intro.pdf

Eschenbach, W. (2016, January 8). *How Thunderstorms Beat the Heat.* Retrieved from Wattsupwiththat.com: https://wattsupwiththat.com/2016/01/08/how-thunderstorms-beat-the-heat/

Eschenbach, W. (2017, May 25). *Estimating Cloud Feedback Using CERES Data.* Retrieved from https://wattsupwiththat.com/2017/05/25/estimating-cloud-feedback-using-ceres-data/

Essex, C., & McKitrick, R. (2003). *Taken By Storm: The Troubled Science, Policy and Politics of Global Warming.* Key Porter Books. Retrieved from https://www.amazon.com/dp/1552632121/?tag=tbs242-20

Evans, W. F., & Puckrin, E. (2006). Measurements of the Radiative Surface Forcing of Climate. *18th Conference on Climate Variability and Change.* Retrieved from https://ams.confex.com/ams/Annual2006/techprogram/paper_100737.htm

ExxonMobil. (2018a). *Cause No. 096-297222-18, Tarrant County District Court, Petitioner's Objections and Response.* Retrieved from https://jnswire.s3.amazonaws.com/jns-media/9b/d2/781207/ExxonStudy.pdf

Fall, S., Watts, A., Nielsen-Gammon, J., Jones, E., Niyogi, D., Christy, J. R., & Sr., R. A. (2011). Analysis of the impacts of station exposure on the U.S. Historical Climatology Network temperatures and temperature trends. *Journal of Geophysical Research, 116*(D14). Retrieved from https://agupubs.onlinelibrary.wiley.com/doi/full/10.1029/2010JD015146

Feynman, J., & Ruzmaikin, A. (2014). The Centennial Gleissberg Cycle and its association with extended minima. *J of Geophysical Research: Space Physics, 119*(8). Retrieved from https://agupubs.onlinelibrary.wiley.com/doi/full/10.1002/2013JA019478

Foster, G., Royer, D., & Lunt, D. (2017). Future climate forcing potentially without precedent in the last 420 million years. *Nature Communications, 8*. Retrieved from https://www.nature.com/articles/ncomms14845

Fröhlich, C. (2003). *Solar Irradiance Variability*. AGU. Retrieved from https://scholar.google.com/scholar?hl=en&as_sdt=0%2C44&q=Fr%C3%B6hlich%2C+C.+Solar+irradiance+variability.+In+Geophysical+Monograph+141%3A+&btnG=

Fröhlich, C., & Lean, J. (1998). The Sun's Total Irradiance: Cycles, Trends and Related Climate Change Uncertainties since 1976. *Geophysical Research Letters, 25*(23). Retrieved from https://agupubs.onlinelibrary.wiley.com/doi/pdf/10.1029/1998GL900157

Fyfe, J., Gillett, N., & Zwiers, F. (2013, September). Overestimated global warming over the past 20 years. *Nature Clim Change*, 767-769. doi:https://doi.org/10.1038/nclimate1972

Ganopolski, A., Winkelmann, R., & Schellnhuber, H. (2016). Critical insolation–CO2 relation for diagnosing past and future glacial inception. *Nature, 529*, 200-203. Retrieved from https://www.nature.com/articles/nature16494

Gehler, A., Gingerich, P., & Pack, A. (2016). Temperature and atmospheric CO2 concentration estimates through the PETM using triple oxygen isotope analysis of mammalian bioapatite. *PNAS, 113*(28). Retrieved from https://www.pnas.org/content/pnas/113/28/7739.full.pdf

Gerhart, L., & Ward, J. (2010). Plant responses to low [CO2] of the past. *New Phytologist, 188*, 674-695. Retrieved from https://nph.onlinelibrary.wiley.com/doi/pdf/10.1111/j.1469-8137.2010.03441.x

Ghienne, J.-F., Heron, D. P., Moreau, J., Denis, M., & Deynoux, M. (2009). The Late Ordovician glacial sedimentary system of the North Gondwana platform. In P. Chistoffersen, & N. Glasser, *Glacial Sedimentary Processes and Products*.

Gillett et al., N. (2013). Constraining the Ratio of Global Warming to Cumulative CO2 Emissions Using CMIP5 Simulations. *J of Climate*, 6844-6858. Retrieved from https://journals.ametsoc.org/view/journals/clim/26/18/jcli-d-12-00476.1.xml?tab_body=pdf

Goldenberg, S. (2015, December 8). Greenpeace exposes sceptics hired to cast doubt on climate science. *The Guardian*. Retrieved from https://www.theguardian.com/environment/2015/dec/08/greenpeace-exposes-sceptics-cast-doubt-climate-science

Goode, P. R., & Palle, E. (2007, September). Shortwave forcing of the Earth's climate: Modern and historical variations in the Sun's irradiance and the Earth's reflectance. *J Atmospheric and Solar-Terrestrial Physics, 69*(13), 1556-1568. Retrieved from https://www.sciencedirect.com/science/article/abs/pii/S1364682607001617

Gouretski, V. (2019). A New Global Ocean Hydrographic Climatology. *Atmospheric and Oceanic Science Letters, 12*(3), 226-229. Retrieved from https://www.tandfonline.com/doi/full/10.1080/16742834.2019.1588066

Government Accounting Office. (2018). *Climate Change: Analysis of Reported Federal Funding.* Washington D.C.: US Government. Retrieved from https://www.gao.gov/products/gao-18-223

Graham, S. (2000, March 24). *Milutin Milankovitch.* Retrieved from NASA Earth Observatory: https://earthobservatory.nasa.gov/features/Milankovitch

Hahne-Waldscheck, B. (2012, May 8). I'm ashamed of that today! *EIKE - European Institute for Climate and Energy.* Retrieved from https://www.eike-klima-energie.eu/2012/05/08/dafuer-schaeme-ich-mich-heute/

Haigh, J. (2011). *Solar Influences on Climate.* Imperial College, London. Retrieved from https://www.imperial.ac.uk/media/imperial-college/grantham-institute/public/publications/briefing-papers/Solar-Influences-on-Climate---Grantham-BP-5.pdf

Haimberger, L., Tavolato, C., & Sperka, S. (2012). Homogenization of the Global Radiosonde Temperature Dataset through Combined Comparison with Reanalysis Background Series and Neighboring Stations. *Journal of Climate,* 8108-8131. Retrieved from https://journals.ametsoc.org/view/journals/clim/25/23/jcli-d-11-00668.1.xml

Hannon, R. (2020, May 27). *The Yin and Yang of Holocent Polar Regions.* Retrieved from andymaypetrophysicist.com: https://andymaypetrophysicist.com/2020/05/27/the-yin-and-yang-of-holocene-polar-regions/

Hansen, B. (2012). *History of Earth's Climate.* Retrieved from https://www.dandebat.dk/eng-klima5.htm

Hansen, K. (2020, February 9). *Sea level: Rise and Fall - Part 5: Bending the trend.* Retrieved from https://wattsupwiththat.com/2020/02/09/sea-level-rise-and-fall-part-5-bending-the-trend/

Happer, W. (2015). *Email Chain between Happer and Greenpeace.* Retrieved from Document Cloud: https://assets.documentcloud.org/documents/2642410/Email-Chain-Happer-O-Keefe-and-Donors-Trust.pdf

Happer, W. (2021a, February 5). *William Happer Interview.* Retrieved from andymaypetrophysicist.com: https://andymaypetrophysicist.files.wordpress.com/2021/03/happer-interview.pdf

Happer, W. (2021b, February 5). *Happer Major Statement.* Retrieved from andymaypetrophysicist.com:

https://andymaypetrophysicist.files.wordpress.com/2021/09/happer_maj or_statement.pdf

Happer, W. (2021c, February 5). *William Happer's detailed response*. Retrieved from andymaypetrophysicist.com: https://andymaypetrophysicist.files.wordpress.com/2021/03/happer_detai led_response.pdf

Happer, W. (2021d, February 5). *William Happer's final reply*. Retrieved from andymaypetrophysicist.com: https://andymaypetrophysicist.files.wordpress.com/2021/04/happers_fina l_reply.pdf

Harde, H. (2014). Advanced Two-Layer Climate Model for the Assessment of Global Warming by CO2. *Open Journal of Atmospheric and Climate Change*. Retrieved from https://www.researchgate.net/profile/Hermann_Harde/publication/2689 81652_Advanced_Two-Layer_Climate_Model_for_the_Assessment_of_Global_Warming_by_CO 2/links/547cbb420cf2cfe203c1fbab.pdf

Hartmann, D., & Michelsen, M. (2002). No Evidence for Iris. *Bull. Amer. Meteor. Soc, 83*, 249-254.

Hausfather, Z., & Peters, G. (2020, January 29). Emissions – the 'business as usual' story is misleading. *Nature, 577*, 618-620. Retrieved from https://www.nature.com/articles/d41586-020-00177-3

Heartland Institute. (2015, April 21). *Willie Soon: Astrophysicist and a Geoscientist Based in Cambridge, Ma*. Retrieved from Heartland.org: https://www.heartland.org/topics/climate-change/willie-soon/index.html

Hoskins, B., & Karoly, D. (1981, June 1). The Steady Linear Response of a Spherical Atmosphere to Thermal and Orographic Forcing. *Journal of Atmospheric Sciences*. Retrieved from https://journals.ametsoc.org/view/journals/atsc/38/6/1520-0469_1981_038_1179_tslroa_2_0_co_2.xml?tab_body=fulltext-display

Hosoda, S., Ohira, T., Sato, K., & Suga, T. (2010). Improved Description of Global Mixed-Layer Depth Using Argo Profiling Floats. *J of Oceanography, 66*, 773-787. Retrieved from https://link.springer.com/article/10.1007/s10872-010-0063-3

Houghton, J. (1996, August 22). Justification of Chapter 8. *Nature, 382*, 665. Retrieved from https://www.nature.com/articles/382665a0.pdf?origin=ppub

Howes, R., & Herzenberg, C. (2003). *Their Day in the Sun: Women of the Manhattan Project (Labor And Social Change)*. Temple University Press. Retrieved from https://www.google.com/books/edition/Their_Day_in_the_Sun/Ys0N4r Fgt6UC?hl=en&gbpv=0

Huang, S. P., Pollack, H. N., & Shen, P.-Y. (2008). A late Quaternary climate reconstruction based on borehole heat flux data, borehole temperature data, and the instrumental record. *Geophys. Res. Let., 35*. Retrieved from https://agupubs.onlinelibrary.wiley.com/doi/full/10.1029/2008GL034187

Works Cited

Huber, M., & Caballero, R. (2011). The Early Eocene equable climate problem revisited. *Climate of the Past, 7,* 603-633. Retrieved from https://cp.copernicus.org/articles/7/603/2011/

Hudson, D. (2017). *Citizen's United v. Federal Election Commission.* Retrieved from The First Amendment Encyclopedia: https://mtsu.edu/first-amendment/article/1504/citizens-united-v-federal-election-commission

Idso, C. (2013). *The Positve Externalities of Carbon Dioxide: Estimating the Monetary Benefits of Rising Atmospheric CO2 Concentrations on Global Food production.* Center for the study of Carbon Dioxide and Global Change. Retrieved from https://www.heartland.org/publications-resources/publications/the-positive-externalities-of-carbon-dioxide-estimating-the-monetary-benefits-of-rising-atmospheric-co2-concentrations-on-global-food-production

Internet Archive. (2021, February 5). *Wayback Machine.* Retrieved from Internet Archive: https://archive.org/web/

IPCC. (1990). *Climate Change, The IPCC Scientific Assessment.* Cambridge: Cambridge University Press. Retrieved from https://www.ipcc.ch/site/assets/uploads/2018/03/ipcc_far_wg_I_full_report.pdf

IPCC. (1996). *SAR, Climate Change 1995, The Science of Climate Change.* Cambridge: Cambridge University Press. Retrieved from https://www.ipcc.ch/site/assets/uploads/2018/02/ipcc_sar_wg_I_full_report.pdf

IPCC. (2001). *Climate Change 2001: The Scientific Basis [TAR].* New York: University Press. Retrieved from https://www.ipcc.ch/site/assets/uploads/2018/03/WGI_TAR_full_report.pdf

IPCC. (2007b). *WG1: Climate Change 2007: The Physical Science Basis (AR4).* Cambridge University Press. Retrieved from https://www.ipcc.ch/site/assets/uploads/2018/05/ar4_wg1_full_report-1.pdf

IPCC. (2013). In T. Stocker, D. Qin, G.-K. Plattner, M. Tignor, S. Allen, J. Boschung, . . . P. Midgley, *Climate Change 2013: The Physical Science Basis. Contribution of Working Group I to the Fifth Assessment Report of the Intergovernmental Panel on Climate Change.* Cambridge: Cambridge University Press. Retrieved from https://www.ipcc.ch/pdf/assessment-report/ar5/wg1/WG1AR5_SPM_FINAL.pdf

IPCC. (2013-WGII_TS). *AR5 WGII Technical Summary.* Retrieved from https://www.ipcc.ch/site/assets/uploads/2018/02/WGIIAR5-TS_FINAL.pdf

IPCC. (2014d). *WGII Summary for Policy Makers.* Retrieved from https://www.ipcc.ch/report/ar5/wg2/summary-for-policymakers/

IPCC. (2018). *Managing the Risks of Extreme Events and Disasters to Advance Climate Change Adaptation.* Retrieved from https://www.ipcc.ch/report/managing-the-risks-of-extreme-events-and-disasters-to-advance-climate-change-adaptation/

IPCC. (2021). Climate Change 2021: The Physical Science Basis. Contribution of Working Group I to the Sixth Assessment Report of the

Intergovernmental Panel on Climate Change. In V. Masson-Delmotte, P. Zhai, A. Pirani, S. L. Connors, C. Péan, S. Berger, . . . B. Zhou (Ed.)., *WG1*. Retrieved from https://www.ipcc.ch/report/ar6/wg1/

IPCC Core writing team. (2010). *Guidance Note for Lead Authors of the IPCC Fifth Assessment Report on Consistent Treatment of Uncertainties.* Retrieved from https://www.ipcc.ch/site/assets/uploads/2017/08/AR5_Uncertainty_Gui dance_Note.pdf

IPCC core writing team. (2014). *Climate Change 2014 Synthesis Report.* Retrieved from https://www.ipcc.ch/report/ar5/syr/

IQ2US. (2007, March 14). *Global Warming is Not a Crisis.* Retrieved from Intelligence Squared Debates: https://www.intelligencesquaredus.org/debates/global-warming-not-crisis

Istvan, R. (2021, January 23). *Overview: 'Ocean Acidification'.* Retrieved from wattsupwiththat.com: https://wattsupwiththat.com/2021/01/23/overview-ocean-acidification/

Jones, N. (2017, January 26). How the World Passed a Carbon Threshold and Why It Matters. *YaleEnvironment360.* Retrieved from https://e360.yale.edu/features/how-the-world-passed-a-carbon-threshold-400ppm-and-why-it-matters

Jones, P. D., New, M., Parker, D. E., Martin, S., & Rigor, I. G. (1999). Surface Air Temperature and its Changes over the Past 150 years. *Reviews of Geophysics, 37*(2), 173-199. Retrieved from https://citeseerx.ist.psu.edu/viewdoc/download?doi=10.1.1.546.7420&rep =rep1&type=pdf

Judge, P., Egeland, R., & Henry, G. (2020, March 1). Sun-like Stars Shed Light on Solar Climate Forcing. *The Astrophysical Journal, 891*(1). Retrieved from https://iopscience.iop.org/article/10.3847/1538-4357/ab72a9/meta

Karl, T. R., Arguez, A., Huang, B., Lawrimore, J. H., Mcmahon, J. R., Menne, M. J., . . . Zhang, H.-M. (2015). Possible artifacts of data biases in the recent global surface warming hiatus. *Science*, 1469-1472. Retrieved from https://science.sciencemag.org/content/348/6242/1469.full

Karl, T., Williams, C., Young, P., & Wendland, W. (1986, February). A Model to Estimate the Time of Observation Bias Associated with Monthly Mean Maximum, Minimum and Mean Temperatures for the United States. *Journal of Climate and Applied Meteorology, 25*(2). Retrieved from https://journals.ametsoc.org/jamc/article/25/2/145/352965/A-Model-to-Estimate-the-Time-of-Observation-Bias

Karoly, D. (1987, November). Southern hemsiphere temperature trends: A possible greenhouse gas effect? *Geophysical Research Letters.* Retrieved from https://agupubs.onlinelibrary.wiley.com/doi/abs/10.1029/GL014i011p01 139

Karoly, D. (1989, May). Northern hemisphere temperature trends: A possible greenhouse gas effect? *Geophysical Research Letters.* Retrieved from https://agupubs.onlinelibrary.wiley.com/doi/abs/10.1029/GL016i005p00 465

Karoly, D. (2021a, February 5). *David Karoly Interview.* Retrieved from andymaypetrophysicist.com:

https://andymaypetrophysicist.files.wordpress.com/2021/03/karoly_inter view-1.pdf

Karoly, D. (2021b, February 5). *Karoly Major Statement.* Retrieved from andymaypetrophysicist.com: https://andymaypetrophysicist.files.wordpress.com/2021/03/karoly_majo r_statement-1.pdf

Keeling, C., Whorf, T., Wong, C. S., & Bellagay, R. (1985). The concentration of atmospheric carbon dioxide at ocean weather station P from 1969 to 1981. *JGR Atmospheres, 90*(D6), 10511-10528. Retrieved from https://agupubs.onlinelibrary.wiley.com/doi/abs/10.1029/JD090iD06p10 511

Keeling, R. F., Graven, H. D., Welp, L. R., Resplandy, L., Bi, J., Piper, S., . . . Meijer, H. (2017). Atmospheric evidence for a global secular increase in carbon isotopic discrimination of land photosynthesis. *PNAS, 114*(39). Retrieved from https://www.pnas.org/content/pnas/114/39/10361.full.pdf

Kennedy, J. J., Rayner, N. A., Smith, R. O., Parker, D. E., & Saunby, M. (2011). Reassessing biases and other uncertainties in sea surface temperature observations measured in situ since 1850; 1. Measurement and sampling uncertainties. *Journal of Geophysical Research, 116.* Retrieved from https://agupubs.onlinelibrary.wiley.com/doi/full/10.1029/2010JD015218

Kennedy, J. J., Rayner, N. A., Smith, R. O., Parker, D. E., & Saunby, M. (2011b). Reassessing biases and other uncertainties in sea surface temperature observations measured in situ since 1850: 2. Biases and homogenization. *J. Geophys. Res., 116.* doi:10.1029/2010JD015220

Kennedy, J., Rayner, N. A., Atkinson, C. P., & Killick, R. E. (2019). An ensemble data set of sea-surface temperature change from 1850: the Met Office Hadley Centre HadSST.4.0.0.0 data set. *JGR Atmospheres, 124*(14). Retrieved from https://agupubs.onlinelibrary.wiley.com/doi/abs/10.1029/2018JD029867

Kirkham, M. B. (2011). *Elevated Carbon Dioxide.* CRC Press. Retrieved from https://www.amazon.com/dp/1439855048/?tag=tbs242-20

Kobashi, T., Kawamura, K., Severinghaus, P., J., . . . E., J. (2011). High variability of Greenland surface temperature over the past 4000 years estimated from trapped air in an ice core. *Geophys. Res. Lett., 38*(21). doi:10.1029/2011GL049444

Kobashi, T., Shindell, D. T., Kodera, K., Box, J., Nakaegawa, T., & Kawamura, K. (2013). On the origin of multidecadal to centennial Greenland temperature anomalies over the past 800 yr. *Climate of the Past, 9*, 583-596. doi:10.5194/cp-9-583-2013

Koonin, S. E. (2021). *Unsettled: What Climate Science Tells us, What it doesn't, and why it matters.* Dallas, Texas, USA: BenBella. Retrieved from https://www.amazon.com/dp/B08JQKQGD5/ref=dp-kindle-redirect?_encoding=UTF8&btkr=1

Kopp, G., & Lean, J. (2011, January 14). A new, lower value of total solar irradiance: Evidence and climate significance. *Geophysical Research Letters, 38*(1). Retrieved from https://agupubs.onlinelibrary.wiley.com/doi/full/10.1029/2010GL045777

Koutsoyiannis, D. (2021, March 19). Rethinking Climate, Climate Change, and Their Relationship with water. *Water, 13*(6). Retrieved from https://www.mdpi.com/2073-4441/13/6/849

Krummheuer, B., & Krivova, N. (2015, July 9). *Solar UV-fluctuations underestimated.* Retrieved from Max Planck Institute: https://www.mps.mpg.de/4017144/PM_2015_07_09_UV-Schwankungen_der_Sonne_unterschaetzt

Lacis, A., Schmidt, G., Rind, D., & Ruedy, R. (2010, October 15). Atmospheric CO2: Principal Control Knob Governing Earth's Temperature. *Science*, 356-359. Retrieved from https://science.sciencemag.org/content/330/6002/356.abstract

Laštovička, J. (2006). Forcing of the ionosphere by waves from below. *Journal of Atmospheric and Solar-Terrestrial Physics, 68*, 479-497. Retrieved from https://www.sciencedirect.com/science/article/abs/pii/S1364682605002579

Leamon, R., & McIntosh, S. (2017, December). *Predicting the La Niña of 2020-21: Termination of Solar Cycles and Correlated Variance in Solar and Atmospheric Variability.* Retrieved from harvard.edu: https://ui.adsabs.harvard.edu/abs/2017AGUFMSH42A..05L/abstract

Leamon, R., McIntosh, S., & Marsh, D. (2020). Termination of Solar Cycles and Correlated Tropospheric Variability. *Earth and Space Science.* Retrieved from https://arxiv.org/abs/1812.02692

Lean, J., Beer, J., & Bradley, R. (1995). Reconstruction of solar irradiance since 1610: Implications for climate change. *Geophysical Research Letters.* Retrieved from https://agupubs.onlinelibrary.wiley.com/doi/abs/10.1029/95GL03093

Lehrer, B. (2007, March 14). *Media Transcripts: Intelligence Squared U.S.: Global Warming is not a Crisis Debate.* Retrieved from Intelligence Squared: https://www.intelligencesquaredus.org/sites/default/files/pdf/transcript-global-warming-is-not-a-crisis.pdf

Lenton, T., Williamson, M., Edwards, N., & al., e. (2006). Millennial timescale carbon cycle and climate change in an efficient Earth system model. *Climate Dynamics, 26*(687). Retrieved from https://link.springer.com/article/10.1007/s00382-006-0109-9#citeas

Lewin, B. (2017). *Searching for the Catastrophe Signal.* Global Warming Policy Foundation. Retrieved from https://www.amazon.com/Searching-Catastrophe-Signal-Origins-Intergovernmental/dp/0993118992

Lewis, N., & Curry, J. (2015). The implications for climate sensitivity of AR5 forcing and heat uptake estimates. *Climate Dynamics, 45*, 1009-1023. Retrieved from https://search.proquest.com/openview/2f4994e4ab3a28571ecdff2edb3aeb13/1?pq-origsite=gscholar&cbl=54165

Lewis, N., & Curry, J. (2018, April 23). The impact of recent forcing and ocean heat uptake data on estimates of climate sensitivity. *Journal of Climate.* Retrieved from https://journals.ametsoc.org/doi/10.1175/JCLI-D-17-0667.1

Lin, B. (2002, June 17). *Evidence Against the Iris Hypothesis.* Retrieved from NASA Earth Observatory: https://earthobservatory.nasa.gov/features/Iris/iris2.php

Lindsey, R. (2003, August 1). *Under a Variable Sun*. Retrieved from Earth Observatory: https://web.archive.org/web/20150908095752/http://earthobservatory.n asa.gov/Features/VariableSun/printall.php

Lindzen, R. (2019). *On Climate Sensitivity*. CO2 Coalition. Retrieved from https://co2coalition.org/wp-content/uploads/2021/08/On-Climate-Sensitivity.pdf

Lindzen, R., & Choi, Y.-S. (2009, August 26). On the determination of climate feedbacks from ERBE data. *Geophysical Research Letters, 36*(16). Retrieved from https://agupubs.onlinelibrary.wiley.com/doi/full/10.1029/2009GL039628

Lindzen, R., & Choi, Y.-S. (2011, August 28). On the Observational Determination of Climate Sensitivity and Implications. *Asia-Pacific Journal of Atmospheric Sciences, 47*(377). Retrieved from https://link.springer.com/article/10.1007/s13143-011-0023-x#citeas

Lindzen, R., & Choi, Y.-S. (2021, April 1). The Iris Effect: A Review. *Asia-pacific Journal of Atmospheric Sciences*. doi:https://doi.org/10.1007/s13143-021-00238-1

Lindzen, R., & Christy, J. (2020). *The Global mean Temperature Anomaly Record: How it works and why it is misleading*. CO2 Coalition. Retrieved from https://co2coalition.org/publications/the-global-mean-temperature-anomaly-record/

Lindzen, R., Chou, M.-D., & Hou, A. (2001, March). Does the Earth have an Adaptive Iris. *Bulletin of the American Meteorological Society, 82*(3). Retrieved from https://journals.ametsoc.org/view/journals/bams/82/3/1520-0477_2001_082_0417_dtehaa_2_3_co_2.xml

Lindzen, R., Chou, M.-D., & Hou, A. (2002). Comment on "No Evidence for Iris". *BAMS, 83*(9), 1345-1349. Retrieved from https://www.jstor.org/stable/26215399?seq=1

Lockwood, G., Skiff, B., Baliunas, S., & Radick, R. (1992, December 17). Long-term solar brightness changes estimated from a survey of Sun-like stars. *Nature, 360*, 653-655. Retrieved from https://www.nature.com/articles/360653a0

Loeb, N. G., Doelling, D., Wang, H., Su, W., Nguyen, C., Corbett, J., & Liang, L. (2018). Clouds and the Earth's Radiant Energy System (CERES) Energy Balanced and Filled (EBAF) Top-of-Atmosphere (TOA) Edition-4.0 Data Product. *Journal of Climate, 31*(2). Retrieved from https://journals.ametsoc.org/view/journals/clim/31/2/jcli-d-17-0208.1.xml

Lomborg, B. (2016, May 5). No one ever says it, but in many ways global warming will be a good thing. *The Telegraph*. Retrieved from https://www.telegraph.co.uk/news/2016/05/05/no-one-ever-says-it-but-in-many-ways-global-warming-will-be-a-go/

Lomborg, B. (2018, October 9). U.N. Ignores Economics of Climate. *Wall Street Journal*. Retrieved from https://www.wsj.com/articles/u-n-ignores-economics-of-climate-1539125496

Lomborg, B. (2020, July). Welfare in the 21st century: Increasing development, reducing inequality, the impact of climate change, and the cost of climate

policies. *Technological Forecasting & Social Change, 156.* Retrieved from
https://www.sciencedirect.com/science/article/pii/S0040162520304157

Luening, S., Vahrenholt, F., & Galka, M. (2020). *Mapping the Medieval Climate Anomaly.*
Retrieved from Google Maps:
https://www.google.com/maps/d/viewer?mid=1akI_yGSUlO_qEvrmrIY
v9kHknq4&ll=-3.81666561775622e-14%2C38.03818700000005&z=1

MacDonald, G. (1981). *The Long Term Impacts of Increasing Atmospheric Carbon Dioxide
Levels.* Ballinger. Retrieved from
https://www.google.com/books/edition/The_Long_term_Impacts_of_In
creasing_Atmo/9wdSAAAAMAAJ?hl=en

Malakoff, D. (2007b, March 22). *Global Warming is not a Crisis.* Retrieved from NPR:
https://www.npr.org/2007/03/22/9082151/global-warming-is-not-a-
crisis

Mann, M. E., Bradley, R. S., & Hughes, M. K. (1998). Global-scale temperature
patterns and climate forcing over the past six centuries. *Nature, 392,* 779-
787. Retrieved from https://www.nature.com/articles/33859

Mann, M., Bradley, R., & Hughes, M. (1999, March 15). Northern Hemisphere
Temperatures during the Past Millennium: Inferences, Uncertainties and
Limitations. *Geophysical Research Letters, 26*(6), 759-762.
doi:https://doi.org/10.1029/1999GL900070

Mann, M., Bradley, R., & Hughes, M. (2004). Correction: Corrigendum: Global-scale
temperature patterns and climate forcing over the past six centuries. *Nature,
430*(105). Retrieved from https://doi.org/10.1038/nature02478

Marcott, S. A., Shakun, J. D., Clark, P. U., & Mix, A. C. (2013, March 8). A
Reconstruction of Regional and Global Temperature for the Past 11,300
Years. *Science,* 1198-1201. Retrieved from
https://science.sciencemag.org/CONTENT/339/6124/1198.abstract

Marsh, N., & Svensmark, H. (2000). Cosmic Rays, Clouds and Climate. *Space Science
Reviews, 94,* 215-230. Retrieved from
https://link.springer.com/article/10.1023/A:1026723423896

Marshall, J. (2018, February 22). *The General Circulation of the Atmosphere.* Retrieved
from Weather Climate Lab: http://weatherclimatelab.mit.edu/wp-
content/uploads/2018/02/General-circulation-of-amosphere.pdf

Mason, J., Painting, R., & Tamblyn, G. (2018). *Introducing Climate Science.* Dunedin
Academic Press. Retrieved from https://www.amazon.com/Introducing-
Climate-Science-Environmental-Sciences/dp/1780460694

Maunder, A., & Maunder, E. W. (1908). *The heavens and their story.* London: Robert
Cully. Retrieved from
https://archive.org/details/heavenstheirstor00maunrich

Mauritsen, T., & Stevens, B. (2015). Missing iris effect as a possible cause of muted
hydrological change and high climate sensitivity in models. *Nature Geoscience,
8,* 346-351. Retrieved from https://www.nature.com/articles/ngeo2414

May, A. (2015b, August 25). *Comparing Early 20th Century Warming to late 20th Century
Warming.* Retrieved from andymaypetrophysicist.com:
https://andymaypetrophysicist.com/comparing-early-20th-century-
warming-to-late-20th-century-warming/

Works Cited

May, A. (2016k, September 27). *Solar Variability and the Earth's Climate*. Retrieved from andymaypetrophysicist.com: https://andymaypetrophysicist.com/solar-variability-and-the-earths-climate/

May, A. (2016p). *The Exxon Climate Papers*. Retrieved from andymaypetrophysicist.com: https://andymaypetrophysicist.com/did-exxon-lie-about-the-dangers-of-climate-change-or-are-they-being-silenced-through-intimidation/

May, A. (2018). *Climate Catastrophe! Science or Science Fiction?* American Freedom Publications LLC. Retrieved from https://www.amazon.com/CLIMATE-CATASTROPHE-Science-Fiction-ebook/dp/B07CPHCBV1/ref=sr_1_1?ie=UTF8&qid=1535627846&sr=8-1&keywords=climate+catastrophe+science+or+science+fiction

May, A. (2018, September 3). *The Great Debate Part D - Summary*. Retrieved from andymaypetrophysicist.com: https://andymaypetrophysicist.com/2018/09/03/the-great-debate-part-d-summary/

May, A. (2018d). *Climate Change: The Great Debate*. The Woodlands, Texas, USA: Andy May. Retrieved from https://andymaypetrophysicist.files.wordpress.com/2018/09/the-great-debate-report1.pdf

May, A. (2018d, June 9). *Does Global Warming increase total atmospheric water vapor (TPW)?* Retrieved from andymaypetrophysicist.com: https://andymaypetrophysicist.com/2018/06/09/does-global-warming-increase-total-atmospheric-water-vapor-tpw/

May, A. (2020c). *Politics and Climate Change: A History*. Springfield, Missouri: American Freedom Publications. Retrieved from https://www.amazon.com/POLITICS-CLIMATE-CHANGE-ANDY-MAY-ebook/dp/B08LJSBVBC/ref=sr_1_1?crid=3POS1QGAQ2C2X&dchild=1&keywords=politics+and+climate+change+a+history+by+andy+may&qid=1609414686&sprefix=Politics+and+Climate%2Caps%2C186&sr=8-1

May, A. (2020d, December 9). *Sea Surface skin temperature*. Retrieved from andymaypetrophysicist.com: https://andymaypetrophysicist.com/2020/12/09/sea-surface-skin-temperature/

May, A. (2020e, November 27). *Ocean Temperature Update*. Retrieved from andymaypetrophysicist.com: https://andymaypetrophysicist.com/2020/11/27/ocean-temperature-update/

May, A. (2021, January 6). *Changing Climate Debate History*. Retrieved from andymaypetrophysicist.com: https://andymaypetrophysicist.com/2021/01/06/changing-climate-debate-history/

McInerney, F., & Wing, S. (2011). The Paleocene-Eocene Thermal Maximum: A Perturbation of Carbon Cycle, Climate, and Biosphere with Implications for the Future. *Annual Review Earth and Planetary Sciences, 39*, 489-516. Retrieved from https://repository.si.edu/bitstream/handle/10088/16827/paleo_McInerney_Wing_2011_AREPS.pdf

McIntyre, S., & McKitrick, R. (2003, November 1). Corrections to the Mann et. al. (1998) Proxy Data Base and Northern Hemispheric Average Temperature Series. *Energy and Environment*. Retrieved from https://journals.sagepub.com/doi/abs/10.1260/095830503322793632

McIntyre, S., & McKitrick, R. (2005). Hockey sticks, principal components, and spurious significance. *Geophysical Research Letters, 32*. doi:10.1029/2004GL021750

McKitrick, R. (2018). *Statement of Ross McKitrick*. Legal Document. Retrieved from http://ielts-yasi.englishlab.net/mckitrick_nyc_lawsuit.pdf

McKitrick, R., & Christy, J. (2018, July 6). A Test of the Tropical 200- to 300-hPa Warming Rate in Climate Models, Earth and Space Science. *Earth and Space Science, 5*(9), 529-536. Retrieved from https://agupubs.onlinelibrary.wiley.com/doi/full/10.1029/2018EA000401

Mears, C., & Wentz, F. (2016, May 15). Sensitivity of Satellite-Derived Tropospheric Temperature Trends to the Diurnal Cycle Adjustment. *Journal of Climate*, 3629-3646. Retrieved from https://journals.ametsoc.org/view/journals/clim/29/10/jcli-d-15-0744.1.xml

Mears, C., Smith, D., Ricciardulli, L., Wang, J., Huelsing, H., & Wentz, F. (2018). Construction and Uncertainty Estimation of a Satellite-Derived Total Precipitable Water Data Record Over the World's Oceans. *Earth and Space Science*. Retrieved from https://agupubs.onlinelibrary.wiley.com/doi/full/10.1002/2018EA000363

Mencken, H. L. (1918). *In Defense of Women*. Retrieved from https://www.amazon.com/Defense-Women-H-L-Mencken-dp-1434116344/dp/1434116344/ref=mt_other?_encoding=UTF8&me=&qid=1598539883

Menne, M., & Williams, C. (2009a). Homogenization of Temperature Series via Pairwise Comparisons. *Journal of Climate, 22*(7), 1700-1717. Retrieved from https://journals.ametsoc.org/jcli/article/22/7/1700/32422

Michaels, P., & Knappenberger, P. (1996). Human effect on global climate? *Nature*, 522-523. Retrieved from https://www.nature.com/articles/384522b0

Miskolczi, F. (2010, August 1). The Stable Stationary Value of the Earth's Global Average Atmospheric Planck-Weighted Greenhouse-Gas Optical Thickness. *Energy and Environment*. Retrieved from http://journals.sagepub.com/doi/abs/10.1260/0958-305X.21.4.243

Montford, A. (2010). *The Hockey Stick Illusion*. Stacey International Publishing. Retrieved from https://www.amazon.com/Hockey-Stick-Illusion-Andrew-Montford-ebook/dp/B0182I73BA/ref=sr_1_1?dchild=1&keywords=The+Hockey+Stick+Illusion&qid=1590414534&sr=8-1

Moran, D. (2017, July 18). *Forget Heat Index. Wet Bulb Globe Temperature Is Where It's At*. Retrieved from DTN: https://www.dtn.com/forget-heat-index-wet-bulb-globe-temperature-is-where-its-at/#:~:text=Wet%20Bulb%20Globe%20Temperature%20%28WBGT%2

Works Cited

9%20is%20similar%20to,affected%20the%20US%20armed%20services%2
0during%20the%201940s.

Morano, M. (2010, August 23). *James Cameron's folly: 'Titanic Chicken of the Sea'*. Retrieved from Watts Up with That: https://wattsupwiththat.com/2010/08/23/james-camerons-folly-titanic-chicken-of-the-sea/

Morice, C. P., Kennedy, J., Rayner, N., Winn, J., Hogan, E., Killick, R., . . . Simpson, I. (2021, Feb. 16). An updated assessment of near-surface temperature change from 1850: the HadCRUT5 dataset. *Journal of Geophysical Research (Atmospheres), 126*(3). Retrieved from https://agupubs.onlinelibrary.wiley.com/doi/full/10.1029/2019JD032361

Moy, C., Seltzer, G., & Rodbell, D. (2002). Variability of El Niño/Southern Oscillation activity at millennial timescales during the Holocene epoch. *Nature, 420*, 162-165. Retrieved from https://doi.org/10.1038/nature01194

Munshi, J. (2018, August 25). Climate Change, Tropospheric Warming, and Stratospheric Cooling. *SSRN*. Retrieved from https://papers.ssrn.com/sol3/papers.cfm?abstract_id=3238535#maincontent

Myers et al., S. (2014). Increasing CO2 threatens human nutrition. *Nature, 510*, 139-142. Retrieved from https://www.nature.com/articles/nature13179?__hstc=107511837.6a5803 56686e0e461a5161b49f40910c.1470787200055.1470787200056.147078720 0057.1&__hssc=107511837.1.1470787200058&__hsfp=1773666937

Nakamura, M. (2018). *Confessions of a climate scientist*. Retrieved from https://www.amazon.com/kikoukagakushanokokuhaku-chikyuuonndannkahamikennshounokasetsu-Japanese-Nakamura-Mototaka-ebook/dp/B07FKHF7T2/ref=cm_cr_arp_d_product_top?ie=UTF8

NASA. (2021). *CERES*. Retrieved from CERES: https://ceres.larc.nasa.gov/science/#ceres-top-of-atmosphere-radiative-flux

NASA. (2021). *Measuring Earth's Albedo*. Retrieved from NASA Earth Observatory: https://earthobservatory.nasa.gov/images/84499/measuring-earths-albedo

National Research Council. (2006). *Surface Temperature Reconstructions for the Last 2,000 Years*. Washington, DC: The National Academies Press. doi:https://doi.org/10.17226/11676

National Weather Service. (2021, March). *Climate Prediction Center*. Retrieved from NOAA: https://origin.cpc.ncep.noaa.gov/products/analysis_monitoring/ensostuff/ONI_v5.php

Newell, R., & Dopplick, T. (1979). Questions Concerning the Possible Influence of Anthropogenic CO2 on Atmospheric Temperature. *J. Applied Meterology, 18*, 822-825. Retrieved from http://journals.ametsoc.org/doi/pdf/10.1175/1520-0450(1979)018%3C0822%3AQCTPIO%3E2.0.CO%3B2

NGRIP team. (2004). High-resolution record of Northern Hemisphere climate extending into the last interglacial period. *Nature, 431*, 147-151. Retrieved from https://www.nature.com/articles/nature02805

Nicholls, N. (1996). An incriminating fingerprint. *Nature, 382*, 27-28. Retrieved from https://www.nature.com/articles/382027a0

NOAA. (2010). *How is ozone formed in the atmosphere*. Retrieved from NOAA.gov: https://csl.noaa.gov/assessments/ozone/2010/twentyquestions/Q2.pdf

NOAA. (2013). *NOAA tides and currents*. Retrieved from Co-Ops Specifications And Deliverables For Installation, Operation, And Removal Of Water Level Station: https://tidesandcurrents.noaa.gov/publications/CO-OPS_Specifications_and_Deliverables_for_installation_operation_and_re moval_of_water_level_stations_updated_October_2013.pdf#:~:text=For %20tidal%20range%20less%20than%20or%20equal%20to,sensor%20resol ution%2

Nordhaus, W. (2018). *Climate change: The Ultimate Challenge for Economics*. Nobel Prize Lecture.

Nussbaumer, S., & Zumbuhl, H. (2012). The Little Ice Age history of the Glacier des Bossons (Mont Blanc massif, France): a new high-resolution glacier length curve based on historical documents. *Climatic Change, 111*, 301-334. doi:10.1007/s10584-011-0130-9

Office of the President. (2013). *Federal Climate Change Expenditures Report to Congress*. US Government. Retrieved from https://obamawhitehouse.archives.gov/sites/default/files/omb/assets/leg islative_reports/fcce-report-to-congress.pdf

Onians, C. (2000, mARCH 20). Snowfalls are now just a thing of the past. *The Independent*. Retrieved from https://wattsupwiththat.com/wp-content/uploads/2015/11/snowfalls-are-now-just-a-thing-of-the-past-the-independent.pdf

Oreskes, N., & Conway, E. (2010). *Merchants of Doubt*. New York: Bloomsbury Press. Retrieved from https://www.google.com/books/edition/Merchants_of_Doubt/fpMh3nh 3JI0C?hl=en&gbpv=1&printsec=frontcover

Orr, J., Fabry, V., & Aumont, O. (2005). nthropogenic ocean acidification over the twenty-first century and its impact on calcifying organisms. *Nature, 437*, 681-686. doi:10.1038/nature04095

Paltridge, G., Arking, A., & Pook, M. (2009). Trends in middle- and upper-level tropospheric humidity from NCEP reanalysis data. *Theor Appl Climatology, 98*, 351-359. Retrieved from https://link.springer.com/article/10.1007/s00704-009-0117-x

Parker, G. (2012). *Global Crisis: War, Climate Change, and Catastrophe in the Seventeenth Century*. Yale University Press. Retrieved from https://www.google.com/books/edition/Global_Crisis/gjdDP15N4FkC? hl=en

Peijnenburg, K. T., Janssen, A. W., Wall-Palmer, D., Goetze, E., Maas, A. E., Todd, J. A., & Marlétaz, F. (2020, October). The origin and diversification of pteropods precede past perturbations in the Earth's carbon cycle. *Proceedings of the National Academy of Sciences, 117*(41), 25609-25617. doi:10.1073/pnas.1920918117

Petit, J., Jouzel, J., Raynaud, D., Barkov, J., Barnola, M., I.Basile, . . . Chappellaz, J. (1999, June 3). Climate and atmospheric history of the past 420,000 years from the Vostok ice core, Antarctica. *Nature, 399*, 429-436. Retrieved from https://cloudfront.escholarship.org/dist/prd/content/qt7rx4413n/qt7rx4413n.pdf

Pielke Jr., R. (2017). *Statement Of Dr. Roger Pielke, Jr. To The Committee On Science, Space, And Technology Of The United States House Of Representatives.* U.S. House of Representatives, Washington, DC. Retrieved from https://science.house.gov/sites/republicans.science.house.gov/files/documents/HHRG-115-SY-WState-RPielke-20170329.pdf

Pielke Jr., R. (2022, January 5). *Global Weather and Climate Disasters 2000 to 2021.* Retrieved from The Honest Broker Newsletter: https://rogerpielkejr.substack.com/p/global-weather-and-climate-disasters

Pielke Jr., R. (2022b, February 2). *U.S. Disaster Costs 1990 to 2019.* Retrieved from The Honest Broker Newsletter: https://rogerpielkejr.substack.com/p/us-disaster-costs-1990-to-2019

Pierrehumbert, R. (2011, January). Infrared radiation and planetary temperature. *Physics Today*, pp. 33-38. Retrieved from http://faculty.washington.edu/dcatling/555_PlanetaryAtmos/Pierrehumbert2011_RadiationPhysToday.pdf

Polissar, P. J., Abbott, M. B., Wolfe, A. P., Vuille, M., & Bezada, M. (2013). Synchronous interhemispheric climate trends. *Proceedings of the National Academy of Sciences, 110*(36), 14551-14556. Retrieved from https://www.pnas.org/content/110/36/14551.short

Polley, H. W., Johnson, H. B., Mayeux, H. S., Brown, D. A., & White, J. W. (1996). Leaf and Plant Water use Efficiency of C4 Species Grown at Glacial to Elevated CO2 Concentrations. *International Journal of Plant Sciences, 157*(2). Retrieved from https://www.journals.uchicago.edu/doi/pdf/10.1086/297335

Popper, K. R. (1962). *Conjectures and Refutations, The Growth of Scientific Knowledge.* New York: Basic Books. Retrieved from http://ninthstreetcenter.org/Popper.pdf

Quinton, P., & MacLeod, K. (2014). Oxygen isotopes from conodont apatite of the midcontinent US: Implications for Late Ordovician climate evolution. *Palaeogeography, Palaeoclimatology, Palaeoecology, 404*, 57-66.

Radanovics, S. (2010). Homogenization of Antarctic Radiosonde Temperature Time Series using ERA-Interim Innovation Statistics. *EGU General Assembly 2010, held 2-7 May, 2010 in Vienna, Austria, p.2645*, (p. 2645). Retrieved from https://ui.adsabs.harvard.edu/abs/2010EGUGA..12.2645R/abstract

Reiny, S. (2016, April 26). *Carbon Dioxide Fertilization Greening Earth, Study Finds.* Retrieved from NASA: https://www.nasa.gov/feature/goddard/2016/carbon-dioxide-fertilization-greening-earth

Ridley, M. (2013b, January 4). How Fossil Fuels Have Greened the Planet. *Wall Street Journal*. Retrieved from https://www.wsj.com/articles/SB10001424127887323374504578217621593679506

Ritchie, J., & Dowlatabadi, H. (2017, December). Why do climate change scenarios return to coal? *Energy, 140*. Retrieved from

https://www.sciencedirect.com/science/article/abs/pii/S03605442173145
97

Rondanelli, R., & Lindzen, R. (2008). Observed variations in convective precipitation fraction and stratiform area with sea surface temperatures. *Journal of Geophysical Research*. Retrieved from https://agupubs.onlinelibrary.wiley.com/doi/epdf/10.1029/2008JD01006
4

Rosenthal, Y., Linsley, B., & Oppo, D. (2013, November 1). Pacific Ocean Heat Content During the Past 10,000 years. *Science*. Retrieved from http://science.sciencemag.org/content/342/6158/617

Rothman, K., Wise, L., & Hatch, E. (2011). Should Graphs of Risk or Rate Ratios be Plotted on a Log Scale? *American Journal of Epidemiology, 174*(3), 376-377. Retrieved from https://academic.oup.com/aje/article/174/3/376/247288

Sage, R. (2003). The evolution of C4 photosynthesis. *New Phytologist*. Retrieved from https://nph.onlinelibrary.wiley.com/doi/full/10.1111/j.1469-8137.2004.00974.x

Santer, B. D., Painter, J. F., Bonfils, C., Mears, C. A., Solomon, S., Wigley, T. M., . . . Wentz, F. J. (2013). Human and natural influences on the changing thermal structure of the atmosphere. *Proceedings of the National Academy of Sciences, 110*(43). Retrieved from https://www.pnas.org/content/110/43/17235.short

Santer, B., Taylor, K., Penner, J., Wigley, T. M., Johns, T. C., Jones, P. D., & Karoly, D. J. (1996a, July 4). A search for human influences on the thermal structure of the atmosphere. *Nature, 382*, 39-46. Retrieved from https://www.nature.com/articles/382039a0

Scafetta, N. (2013, November). Discussion on climate oscillations: CMIP5 general circulation models versus a semi-empirical harmonic model based on astronomical cycles. *Earth Science Reviews, 126*, 321-357. Retrieved from https://www.sciencedirect.com/science/article/abs/pii/S00128252130014
02

Scafetta, N. (2021, January 17). Detection of non-climatic biases in land surface temperature records by comparing climatic data and their model simulations. *Climate Dynamics*. Retrieved from https://doi.org/10.1007/s00382-021-05626-x

Scafetta, N., & Vahrenholt, F. (2022-in press). Why does the IPCC downplay the Sun? In A. M. Marcel Crok, *CLINTEL's Assessment of AR6*. The Netherlands: Clintel.org. Retrieved from https://clintel.org/

Scafetta, N., & Willson, R. (2014). ACRIM total solar irradiance satellite composite validation versus TSI proxy models. *Astrophysics and Space Science, 350*(2), 421-442. Retrieved from https://link.springer.com/article/10.1007/s10509-013-1775-9

Scafetta, N., Willson, R., Lee, J., & Wu, D. L. (2019). Modeling Quiet Solar Luminosity Variability from TSI Satellite Measurements and proxy Models during 1980-2018. *Remote Sensing, 11*. Retrieved from https://www.mdpi.com/2072-4292/11/21/2569

Schmidtko, S., Johnson, G., & Lyman, J. (n.d.). MIMOC: A global monthly isopycnal upper-ocean climatology with mixed layers. *Journal of Geophysical*

Research: Oceans, 118, 1658-1672. Retrieved from https://agupubs.onlinelibrary.wiley.com/doi/full/10.1002/jgrc.20122

Schmitt, R. (2018). The Ocean's Role in Climate. *Oceanography, 31*(2), 32-40. Retrieved from https://www.jstor.org/stable/26542649

Schmitz, B., Farley, K. A., Goderis, S., Heck, P. R., Bergström, S. M., Boschi, S., . . . Martin, E. (2019, Sep 18). An extraterrestrial trigger for the mid-Ordovician ice age: Dust from the breakup of the L-chondrite parent body. *Science Advances, 5*(9). Retrieved from https://advances.sciencemag.org/content/5/9/eaax4184

Schwartz, J. (2015, December 9). Greenpeace Subterfuge Tests Climate Research. *New York Times.* Retrieved from https://www.nytimes.com/2015/12/09/science/greenpeace-subterfuge-tests-climate-research.html

Scotese, C. (2015). *Some thoughts on Global Climate Change: The Transition from Icehouse to Hothouse.* PALEOMAP Project. Retrieved from https://andymaypetrophysicist.files.wordpress.com/2021/07/scotese_some-thoughts-on-global-climate-changev21.pdf

Scotese, C., Song, H., Mills, B. J., & Meer, D. v. (2021, January). Invited Review Phanerozoic Paleotemperatures: The Earth's Changing Climate during the Last 540 Million Years. *Earth-Science Reviews.* Retrieved from https://www.sciencedirect.com/science/article/abs/pii/S0012825221000027

Scott, M., & Lindsey, R. (2020, June 18). *What's the hottest Earth's ever been?* Retrieved from NOAA Climate.gov: https://www.climate.gov/news-features/climate-qa/whats-hottest-earths-ever-been

Scripps Institute of Oceanography. (2021, January 4). *Scripps CO2 Program.* Retrieved from Scripps: https://scrippsco2.ucsd.edu/#:~:text=CO2%20Concentration%20at%20Mauna,being%20continued%20by%20Ralph%20F.

Seidel, D. J., Zhang, Y., Beljaars, A., Golaz, J.-C., Jacobson, A. R., & Medeiros, a. B. (2012). Climatology of the planetary boundary layer over the continental United States and Europe. *J. Geophys. Res. Atmos., 117.* Retrieved from https://agupubs.onlinelibrary.wiley.com/doi/full/10.1029/2012JD018143

Seitz, F. (1996, June 12). A Major Deception on Global Warming. *Wall Street Journal.* Retrieved from https://www.wsj.com/articles/SB834512411338954000

Severinghaus, J. P., Sowers, T., Brook, E. J., Alley, R. B., & Bender, M. L. (1998, January 8). Timing of abrupt climate change at the end of the Younger Dryas interval from thermally fractionated gases in polar ice. *Nature, 391,* pp. 141-146. Retrieved from http://shadow.eas.gatech.edu/~jean/paleo/Severinghaus_1998.pdf

Shapiro, A., Schmutz, W., Rozanov, E., Schoell, M., Haberreiter, M., Shapiro, A. V., & Nyeki, S. (2011, May). A new approach to the long-term reconstruction of the solar irradiance leads to large historical solar forcing. *Astronomy and Astrophysics, 529.* Retrieved from https://www.aanda.org/index.php?Itemid=129&access=doi&doi=10.1051/0004-6361/201016173&option=com_article

Shaviv, N. (2003b). The Spiral structure of the Milky Way, cosmic rays, and ice age epochs on Earth. *New Astronomy.* Retrieved from http://old.phys.huji.ac.il/~shaviv/articles/long-ice.pdf

Shaviv, N. (2008). Using the oceans as a calorimeter to quantify the solar radiative forcing. *Journal of Geophysical Research, Space Physics, 113*(A11). Retrieved from https://agupubs.onlinelibrary.wiley.com/doi/full/10.1029/2007JA012989

Shaviv, N., & Veizer, J. (2003, July). Celestial driver of Phanerozoic Climate? *GSA Today.* Retrieved from https://www.geosociety.org/gsatoday/archive/13/7/pdf/i1052-5173-13-7-4.pdf

Sheehan, P. (2001). The Late Ordovician Mass Extinction. In *Annual Review of Earth And Planetary Sciences* (Vol. 29, pp. 331-364). Retrieved from https://ui.adsabs.harvard.edu/abs/2001AREPS..29..331S/abstract

Sherwood, S. C., Webb, M. J., Annan, J. D., Armour, K. C., J., P. M., Hargreaves, C., . . . Knutti, R. (2020, July 22). An Assessment of Earth's Climate Sensitivity Using Multiple Lines of Evidence. *Reviews of Geophysics, 58.* doi:https://doi.org/10.1029/2019RG000678

Sherwood, S., & Huber, M. (2010). An Adaptability limit to climate change due to heat stress. *PNAS, 107*(21). Retrieved from https://www.pnas.org/content/107/21/9552.short

Shulman, S. (2012). *Establishing Accountability for Climate Change Damages: Lessons from Tobacco Control.* La Jolla: Climate Accountablity Institute and the Union of Concerned Scientists. Retrieved from https://climateaccountability.org/pdf/Climate%20Accountability%20Rpt%20Oct12.pdf

SILSO. (2020, August 10). *Sunspot Number.* Retrieved from Sunspot Index and Long-term Solar Observations: http://www.sidc.be/silso/datafiles

Soden, B. J., Wetherald, R., Stenchikov, G., & Robock, A. (2002). Global Cooling After the Eruption of Mount Pinatubo: A Test of Climate Feedback by Water Vapor. *Science, 296*(5568), 727-730. Retrieved from http://science.sciencemag.org/content/296/5568/727

Soon, W., & Baliunas, S. (2003). Proxy climatic and environmental changes of the past 1000 years. *Climate Research*, 89-110. Retrieved from https://www.int-res.com/abstracts/cr/v23/n2/p89-110/

Soon, W., Connolly, R., & Connolly, M. (2015). Re-evaluating the role of solar variability on Northern Hemisphere temperature trends since the 19th century. *Earth Science Reviews, 150*, 409-452. Retrieved from https://www.sciencedirect.com/science/article/pii/S0012825215300349

Soon, W., Connolly, R., Connolly, M., O'Neill, P., Zheng, J., Ge, Q., . . . Yan, H. (2018). Comparing the Current and Early 20th Century Warm Periods in China. *Earth-Science Reviews*, 80-101. Retrieved from https://www.sciencedirect.com/science/article/abs/pii/S0012825218301235

Spencer, R. (2021, February 2). *UAH Global Temperature Update for January 2021.* Retrieved from drroyspencer.com: http://www.drroyspencer.com/2021/02/uah-global-temperature-update-for-january-2021-0-12-deg-c-new-base-period/

Works Cited

Spencer, R., & Braswell, W. (1997). How dry is the tropical free troposphere? Implications for global warming theory. *Bull. Am. Meteor. Soc., 78*(6). Retrieved from https://journals.ametsoc.org/view/journals/bams/78/6/1520-0477_1997_078_1097_hdittf_2_0_co_2.xml

Spencer, R., & Christy, J. (1990). Precise Monitoring of Global Temperature Trends from Satellites. *Science, 247*. Retrieved from https://science.sciencemag.org/content/247/4950/1558.abstract

Springer, M., Murphy, W., Eizirik, E., & O'Brien, S. (2003, February). Placental mammal diversification and the Cretaceous-Tertiary Boundary. *PNAS, 100*(3). Retrieved from https://www.pnas.org/content/pnas/100/3/1056.full.pdf

Steele, J. (2017, March 1). *How NOAA and Bad Modeling Invented an "Ocean Acidification" Icon: Part 1 – Sea Butterflies.* Retrieved from wattsupwiththat.com: https://wattsupwiththat.com/2017/03/01/how-noaa-and-bad-modeling-invented-an-ocean-acidification-icon-part-1-sea-butterflies/

Steele, J. (2017b, March 2). *How NOAA and Bad Modeling Invented an "Ocean Acidification" Icon: Part 2 – Bad Models.* Retrieved from Wattsupwiththat.com: https://wattsupwiththat.com/2017/03/02/how-noaa-and-bad-modeling-invented-an-ocean-acidification-icon-part-2-bad-models/

Stokke, E. W., Jones, M. T., Tierney, J. E., Svensen, H. H., & Whiteside, J. H. (2020). Temperature changes across the Paleocene-Eocene Thermal Maximum – a new high-resolution TEX86 temperature record from the Eastern North Sea Basin. *Earth and Planetary Science Letters, 544*. Retrieved from https://www.sciencedirect.com/science/article/pii/S0012821X20303320

Stuiver, M., & Quay, P. (1980, January 4). Changes in Atmospheric Carbon-14 Attributed to a Variable Sun. *Science*, 11-19. Retrieved from https://www.science.org/doi/abs/10.1126/science.207.4426.11

Sun, D., & Lindzen, R. (1993). Distribution of tropical tropospheric water. *J. Atmos. Sci., 50*(12). Retrieved from https://journals.ametsoc.org/view/journals/atsc/50/12/1520-0469_1993_050_1643_dottwv_2_0_co_2.xml

Svensmark, H. (1998, November 30). Influence of Cosmic Rays on Earth's Climate. *Physical Review Letters, 81*. Retrieved from https://journals.aps.org/prl/abstract/10.1103/PhysRevLett.81.5027

Svensmark, H. (2019). *Force Majeure, The Sun's role in climate change.* GWPF. Retrieved from https://www.thegwpf.org/content/uploads/2019/03/SvensmarkSolar2019-1.pdf

Tamblyn, G. (2021, February 5). *Glenn Tamblyn.* Retrieved from skepticalscience.com: https://skepticalscience.com/posts.php?u=2178

Tamblyn, G. (2021a, February 5). *Glenn Tamblyn's detailed response.* Retrieved from andymaypetrophysicist.com: https://andymaypetrophysicist.files.wordpress.com/2021/04/glenn-tamblyn-detailed-response.pdf

Tamblyn, G. (2021b, February 5). *Glenn Tamblyn's final reply*. Retrieved from andymaypetrophysicist.com: https://andymaypetrophysicist.files.wordpress.com/2021/03/tamblyn_final_reply.pdf

Taub, D. (2010). Effects of Rising Atmospheric Concentrations of Carbon Dioxide on Plants. *Nature Education, 3*(10). Retrieved from https://www.nature.com/scitable/knowledge/library/effects-of-rising-atmospheric-concentrations-of-carbon-13254108/

Taub, D., & Wang, X. (2008, October 23). Why are Nitrogen Concentrations in Plant Tissues Lower under Elevated CO2? A Critical Examination of the Hypotheses. *JIPB*. Retrieved from https://onlinelibrary.wiley.com/doi/full/10.1111/j.1744-7909.2008.00754.x

Taylor, J. (2011, September 28). The Global Warming Debate Produces An Indisputable Winner. *Forbes*. Retrieved from https://www.forbes.com/sites/jamestaylor/2011/09/28/the-global-warming-debate-produces-an-indisputable-winner/?sh=62fe4a76668f

The Nobel Prize. (2018). *William D. Nordhaus*. Retrieved from The Nobel Prize: https://www.nobelprize.org/prizes/economic-sciences/2018/nordhaus/facts/

TheBestSchools.org. (2021b, February 5). *Introduction to the debate*. Retrieved from andymaypetropysicist.com: https://andymaypetrophysicist.files.wordpress.com/2021/03/introduction_to_debate.pdf

Tissue, D., Griffin, K. L., Thomas, R. B., & Strain, B. R. (1994, August). Effects of low and elevated CO2 on C3 and C4 annuals II. *Oecologia*. Retrieved from https://link.springer.com/article/10.1007/BF00328895

Trenberth, K. (1991). Climate Change and Climate Variability: The Climate Record. In *Managing Water Resources in the West Under Conditions of Climate Uncertainty* (p. 358). Retrieved from https://www.nap.edu/download/1911

Tripathi, R. (2016). *Vada in Theory and Practice*. D. K. Printworld Pvt. Ltd. Retrieved from https://www.exoticindiaart.com/book/details/vada-in-theory-and-practice-studies-in-debates-dialogues-and-discussions-in-indian-intellectual-discourses-NAN415/

Twain, M. (1883). *Life on the Mississippi*. Boston: James R. Osgood and Company. Retrieved from https://www.gutenberg.org/files/245/245-h/245-h.htm#linkc17

Verkley, W. T., & Gerkema, T. (2004). On Maximum Entropy Profiles. *J of the Atmospheric Sciences, 61*(8). Retrieved from https://journals.ametsoc.org/view/journals/atsc/61/8/1520-0469_2004_061_0931_omep_2.0.co_2.xml

Vinós, J. (2016, October 24). *Nature Unbound I: The Glacial Cycle*. Retrieved from judithcurry.com: https://judithcurry.com/2016/10/24/nature-unbound-i-the-glacial-cycle/

Vinós, J. (2016, May 11). *Periodicities in solar variability and climate change: A simple model*. Retrieved from Energy Matters: http://euanmearns.com/periodicities-in-solar-variability-and-climate-change-a-simple-model/

Vinós, J. (2017, April 30). *Nature Unbound III: Holocene climate variability (Part A)*. Retrieved from Climate Etc.: https://judithcurry.com/2017/04/30/nature-unbound-iii-holocene-climate-variability-part-a/

Vinós, J. (2018, August 14). *Nature Unbound X - The Next Glaciation*. Retrieved from Climate, Etc.: https://judithcurry.com/2018/08/14/nature-unbound-x-the-next-glaciation/

Wang, J., Feng, L., Tang, X., Bentley, Y., & Höök, M. (2017). The implications of fossil fuel supply constraints on climate change projections - A Supply Side analysis. *Science Direct, 86*, 58-72. Retrieved from https://www.sciencedirect.com/science/article/abs/pii/S0016328715300690

Wang, S., Ju, W., Penuelas, J., Cescatti, A., & Zhou, Y. (2019, July). Urban-rural gradients reveal joint control of elevated CO_2 and temperature on extended photosynthetic seasons. *Nature Ecology and Evolution, 3*, 1076-1085. Retrieved from https://www.nature.com/articles/s41559-019-0931-1

Wang, Y. M., Lean, J., & Sheeley, N. R. (2005, May 20). Modeling the Sun's Magnetic Field and Irradiance since 1713. *The Astronomical Journal, 625*, 522-538. Retrieved from https://iopscience.iop.org/article/10.1086/429689/pdf

Ward, J. (2005). Evolution and Growth of Plants in a Low CO_2 World. In J. Ehleringer, T. Cerling, & D. Dearing, *A History of Atmospheric CO_2, and its effects on Plants, Animals and ecosystems* (pp. 232-257). New York: Springer. Retrieved from https://www.springer.com/us/book/9780387220697?utm_campaign=bookpage_about_buyonpublisherssite&utm_medium=referral&utm_source=springerlink

Ward, J., Harris, J., Cerling, T., Wiedenhoeft, A., Lott, M., Dearing, M.-D., . . . Ehleringer, J. (2005, January 18). Carbon starvation in glacial trees recovered from the La Brea tar pits, southern California. *PNAS, 102*(3), 690-694. Retrieved from https://www.pnas.org/content/102/3/690.short

Watts, A. (2016b, March 2). *The 'Karlization' of global temperature continues – this time RSS makes a massive upwards adjustment*. Retrieved from Wattsupwiththat.com: https://wattsupwiththat.com/2016/03/02/the-karlization-of-global-temperature-continues-this-time-rss-makes-a-massive-upwards-adjustment/

Watts, A. (2018, August 28). *Climate activists have long history of ducking debates with skeptics*. Retrieved from Watts Up With That?: https://wattsupwiththat.com/2018/08/28/climate-activists-have-long-history-of-ducking-debates-with-skeptics/

Wegman, E., Scott, D., & Said, Y. (2010). *Ad Hoc Committee Report on the Hockey Stick Global Climate Reconstruction*. U.S. Congress. Science and Public Policy Institute. Retrieved from http://scienceandpublicpolicy.org/wp-content/uploads/2010/07/ad_hoc_report.pdf

Wijngaarden, W., & Happer, W. (2020, June 4). Dependence of Earth's Thermal Radiation on Five Most Abundant Greenhouse Gases. *arXiv*. Retrieved from https://arxiv.org/abs/2006.03098

Williams, I., & Pierrehumbert, R. (2017). Observational evidence against strongly stabilizing tropical cloud feedbacks. *Geophysical Research Letters, 44*(3). Retrieved from

https://agupubs.onlinelibrary.wiley.com/doi/epdf/10.1002/2016GL07220
2

Williams, J.-P., Paige, D., Greenhagen, B., & Sefton-Nash, E. (2017, February). The global surface temperatures of the Moon as measured by the Diviner Lunar Radiometer Experiment. *Icarus, 283*, 300-325. Retrieved from https://www.sciencedirect.com/science/article/pii/S0019103516304869

Willoughby, R. (2021, May 23). *Ocean Surface Temperature Limit-Part 1*. Retrieved from Wattsupwiththat: https://wattsupwiththat.com/2021/05/23/ocean-surface-temperature-limit-part-1/

Willson, R. (1997). Total Solar Irradiance Trend During Solar Cycles 21 and 22. *Science, 277*(5334), 1963-1965. Retrieved from https://science.sciencemag.org/content/277/5334/1963

Willson, R. (2014). ACRIM3 and the Total Solar Irradiance database. *Astrophys Space Sci, 352*, 341-352. Retrieved from https://link.springer.com/article/10.1007/s10509-014-1961-4

World Climate Research Programme. (2021, February 24). *WCRP Coupled Model Intercomparison Project (CMIP)*. Retrieved from WCRP: https://www.wcrp-climate.org/wgcm-cmip

World Meteorological Organization. (2021, February 8). *Starting Point*. Retrieved from KNMI Climate Explorer: https://climexp.knmi.nl/start.cgi

Wu, T., Hu, A., Gao, F., Zhang, J., & Meehl, G. (2019). New insights into natural variability and anthropogenic forcing of global/regional climate evolution. *Nature Partner Journals, 2*(18). doi:https://doi.org/10.1038/s41612-019-0075-7

Wyatt, M., & Curry, J. (2014, May). Role for Eurasian Arctic shelf sea ice in a secularly varying hemispheric climate signal during the 20th century. *Climate Dynamics, 42*(9-10), 2763-2782. Retrieved from https://link.springer.com/article/10.1007/s00382-013-1950-2#page-1

Zelinka, M., & Hartmann, D. (2011). The observed sensitivity of high clouds to mean surface temperature anomalies in the tropics. *Journal of Geophysical Research Atmospheres, 116*(D23). Retrieved from https://agupubs.onlinelibrary.wiley.com/doi/full/10.1029/2011JD016459

Zhang, F., Romaniello, S. J., Algeo, T. J., Lau, K. V., Clapham, M. E., Richoz, S., . . . Anbar, A. D. (2018, April 11). Multiple episodes of extensive marine anoxia linked to global warming and continental weathering following the latest Permian mass extinction. *Science Advances, 4*(4). Retrieved from https://advances.sciencemag.org/content/4/4/e1602921.short

Zhu, J., Poulsen, C., & Tierney, J. (2019). Simulation of Eocene extreme warmth and high climate sensitivity through cloud feedbacks. *Science Advances, 5*(9). Retrieved from https://advances.sciencemag.org/content/5/9/eaax1874

Zhu, Z., Piao, S., & Myneni, R. (2016). Greening of the Earth and its drivers. *Nature Clim Change, 6*, 791-795. Retrieved from https://www.nature.com/articles/nclimate3004

Table of Figures

Index

J

K

L

M

U

About the Author

Andy May is a writer, blogger and author living in The Woodlands, Texas. He was born in Lawrence, Kansas, but never really appreciated how interesting Kansas history was until he researched his second book. He enjoys golf and traveling in his spare time. He is also an editor for the climate blog Wattsupwiththat.com, where he has published numerous posts and is the author or co-author of seven peer-reviewed papers on various geological, engineering and petrophysical topics. He has also written about computers and computer software. His personal blog is andymaypetrophysicist.com.

He retired from a 42-year career in petrophysics in 2016. Most of his petrophysical work was for several oil and gas companies worldwide. He has worked in exploring, appraising, and developing oil and gas fields in the U.S., Argentina, Brazil, Indonesia, Thailand, China, the U.K. North Sea, Canada, Mexico, Venezuela, and Russia. He helped discover and appraise several large oil and gas fields.

Late in his career, he worked on unconventional shale oil and gas petrophysics and developed many unique techniques for evaluating these difficult reservoirs. In cooperation with Professor Mike Lovell (University of Leicester in the U.K.) he developed a one-week course in shale reservoir petrophysics. Andy has a B.S. in Geology from the University of Kansas

9 781639 446766